Health, safety and ergonomics

Health, safety and ergonomics

Edited by

Andrew S. Nicholson,
BSc(Hons), MErgS, MIOSH
Research Fellow, Ergonomics Research Unit, Robens Institute of Industrial and Environmental Health and Safety, University of Surrey

and

John E. Ridd,
FIMatM, MEergS, MIOSH
Deputy Head, Ergonomics Research Unit, Robens Instute of Industrial and Environmental Health and Safety, University of Surrey

Butterworths
London   Boston   Singapore   Sydney   Toronto   Wellington

First published 1988

© **Butterworth & Co. (Publishers) Ltd, 1988**

British Library Cataloguing in Publication Data

Health, safety and ergonomics.
1. Ergonomics
I. Nicholson, Andrew S. II. Ridd, John E.
620.8′2

ISBN 0-408-02386-4

Library of Congress Cataloging in Publication Data

Health, Safety, and ergonomics.

  Proceedings of a symposium held Oct. 1, 1987.
  Includes bibliographies and index.
  1. Human engineering—Congresses.
2. Industrial safety—Congresses.
3. Industrial hygiene—Congresses.
I. Nicholson, Andrew S.   II. Ridd, John E.
TA166.H34 1988   620.8′2   88-22238
ISBN 0-408-02386-4

Photoset by Butterworths Litho Preparation Department
Printed and bound in Great Britain by Anchor Brendon Ltd, Tiptree, Essex

Preface

Work and the workplace have been recognized for many years as having a bearing on the health and safety of the worker and, in many cases, also on that of the public at large. Legislation has been introduced in the UK, as in many other countries, to improve the levels of health and safety. However, in time of industrial economic stringency, such as that which now seems to have arisen, the emphasis has been on maintaining jobs and industrial solvency; whether this approach has been successful is not for discussion here, but inevitably health and safety programmes have suffered.

It is perhaps reasonable that any proposals directed at improving the levels of health and safety have to be supported within this strict financial climate by sound economic argument. The financial penalties of deficiencies in health and safety practice are usually borne by the corporate body and rarely by individual departments, they are, therefore, not always immediately apparent. Hence, the introduction of corrective measures on the 'shopfloor', where any implementation costs would normally be the responsibility of the department, is greeted with little enthusiasm.

It is clearly necessary to provide realistic estimates of the benefits that are to be accrued from such actions and to present these in terms that can be appreciated at shopfloor level in order to gain acceptance for health and safety improvements. Yet clear and irrefutable arguments in support of health and safety improvements may be difficult to present in economic terms alone. However, the application of ergonomics to work processes can both improve health and safety standards and increase work efficiency at the same time, thus creating the potential for raised output and increased production – aspects of current industrial life that are much more likely to receive a sympathetic hearing than the apparently non-productive areas of health and safety alone.

Despite the numerous benefits of ergonomics (which appear to be self-evident to the 'converted') the application of this science within industrial and commerical organizations is not commonly undertaken; even where it is utilized it is often as a corrective procedure when a work practice has been found to be inadequate. In those situations where successful ergonomic change has been implemented it is not unusual, with the benefit of hindsight, for such changes to be dismissed as having been 'common sense', and for their introduction deemed to have required little skill or effort. The skill, however, is in the identification of the need for change and in the design of the manner of change such that it will be effective.

It is rare for health and safety professionals to have enough time to become sufficiently skilled in ergonomics techniques for specific workplace evaluations at anything more than a superficial level, even though it is possible to improve an individual's ability to identify potentially hazardous situations. Nevertheless, this is itself a worthwhile activity and it therefore seemed appropriate to bring the advantages, and the methods, of good ergonomics practice to the attention of those most in need of their application. The international symposium entitled 'Workwise Ergonomics, Health and Safety' was born of this concern and the chapters of this book largely constitute the *Proceedings* of that meeting, held on 1 October 1987. The aims of the symposium, and to an even greater degree the aims of this book, are principally to show that:

1. Ergonomics has a direct and important part to play in the prevention of occupational ill health and accidental injury at work.
2. Ergonomics can be used as an aid to management decision-making on the design, construction and operation of plant and equipment, and in the preparation of work systems such that when commissioned, such systems and apparatus will function more efficiently.
3. The quest for improving levels of occupational health and safety, utilizing ergonomic principles, can bring wider benefits by maximizing performance, comfort and worker morale.

In order to make this book a more complete account of the part ergonomics has to play in the improvement of working conditions, further information has been incorporated into a number of the papers and three extra chapters have been added.

The reader will find that some topics are addressed in more than one chapter but that any overlap is complementary and necessary if the section were to be read in isolation. We trust that the editing of the chapters has resulted in some measure of uniformity of presentation and that it is a readable, if extensive, account of the manner in which ergonomics principles may be applied to any work situation to improve working conditions.

In the first volume of his *History of England,* Macaulay wrote, concerning industrial working conditions in 1685:

> The more carefully we examine the history of the past, the more reason we shall find to dissent from those who imagine that our age has been fruitful of new social evils. The truth is that evils are, with scarcely an exception, old. That which is new is the intelligence which discerns and the humanity which remedies them.

The problems of work and workplace design are addressed in this book and in this context ergonomics may reasonably claim to be that intelligence and that humanity; the editors hope that the following contributions will assist those at the 'sharp end' to apply these attributes to the benefit of all.

John E. Ridd and Andrew S. Nicholson
Robens Institute of Industrial and Environmental Health and
Safety, December, 1987

Acknowledgements

The editors wish to express their thanks to the contributors who, at times under duress from various sources, managed on the whole to respond to the inevitable deadlines. We also express our thanks to our colleagues in the Ergonomics Research Unit of the Robens Institute who supported this venture with their encouragement and interest. Various staff of the Health and Safety Executive were involved in the early stages of planning the symposium which provided the stimulus for this book. Their support is gratefully acknowledged. The staff of Butterworth Scientific Ltd were most helpful in seeing this project through to its completion. Finally, we express our thanks to our wives and families who have so patiently supported us.

Contributors

Peter Buckle, BSc, MSc, PhD, FErgS
Lecturer and Deputy Head, Ergonomics Research Unit. Robens Institute of Industrial and Environmental Health and Safety, University of Surrey, UK

E. N. Corlett, DSc, PhD, FEng, FErgS, ABPsS
Department of Production Engineering and Production Management, University of Nottingham, UK

E. J. Cullen, FEng, PhD
Chairman, Health and Safety Commission; President-Elect, Institution of Chemical Engineers, 1988/89 Health and Safety Commission, Baynards House, 1 Chepstow Place, Westbourne Grove, London, UK

P. R. Davis, PhD, MB, BS, FRCS, FErgS
Polvean, Freshwater Lane, St Mawes, Cornwall, UK

D. E. Embrey, BSc, PhD
Managing Director, Human Reliability Associates Ltd, 1 School House Lane, Higher Lane, Dalton, Wigan, UK

John G. Fox
Directorate Health and Safety, Commission of the European Communities, Bâtiment Jean Monnet, Luxembourg

Geoffrey Laycock, BSc(Hons), MIOSH
Formerly HM Inspector of Factories; Personnel Services Manager, Express Dairy Ltd, South Ruislip, Middlesex, UK

Andrew Nicholson, BSc(Hons), MErgS, MIOSH
Research Fellow, Ergonomics Research Unit, Robens Institute of Industrial and Environmental Health and Safety, University of Surrey, UK

Stephen Pheasant, MA, MSc, PhD, FErgS
Lecturer in anatomy, Royal Free Hospital School of Medicine, London, UK; Honorary Consultant at the Robens Institute of Industrial and Environmental Health and Safety, University of Surrey, UK

John E. Ridd, FIMatM, MErgS, MIOSH
Deputy Head, Ergonomics Research Unit, Robens Institute of Industrial and Environmental Health and Safety, University of Surrey, UK

Geoff Simpson, BSc(Tech), MSc, FBPsS, FErgS, MIOSH
Head of Ergonomics Branch of the Institute of Occupational Medicine, Edinburgh, UK

N. T. Thomas, MPhil, FErgS
Senior Lecturer, Ergonomics Unit, Department of Mechanical and Production Engineering, The Polytechnic of Wales, UK

Contents

Part 4   Perspectives on current issues

Part 1

The role of ergonomics

Chapter 1

The advantages of ergonomics intervention

E. J. Cullen

This opening chapter does not address a specific topic but introduces an hypothesis that the succeeding chapters set out to consider. The hypothesis presented is that: 'The application of ergonomics within any and every industry or business can have benefits both for the worker and for the work process in a number of different ways, relating to health, safety and efficiency.' The text offered here is an edited version of the opening address given by Dr John Cullen, Chairman of the Health and Safety Commission, at the international symposium entitled 'Workwise: Ergonomics, Health and Safety'.

The importance of ergonomics in the prevention of accidents and ill health at work has been recognized for some time. It is now appropriate to bring together many of the diverse areas of ergonomics for discussion in relation to the improvement of occupational health, safety and efficiency.

Active and extensive co-operation between the Robens Institute of Industrial and Environmental Health and Safety, the Health and Safety Executive (HSE) and the Ergonomics Society led to the 'Workwise' symposium in October 1987 and to the subsequent publication of extended presentations in this book.

I am sure that the following chapters will enable the reader to gain a better understanding of the link between occupational ill health, accidents and the manner in which ergonomics may have a direct bearing on their occurrence. This will be of benefit to management decision-making in the design, installation and operation of plant and systems of work and also to the workers who are daily exposed to the effects of poor ergonomic design.

It is important to recognize that any ergonomics principles applied to health and safety can also result in other benefits such as increased efficiency and better worker morale.

In this opening chapter, I wish to take the opportunity to consider three aspects of ergonomics and health and safety at work: (1) to discuss the link between ergonomics and health and safety, and suggest that it affects all our work activities; (2) to

,consider how ergonomics principles may be applied at the workplace; and (3) to outline two areas where the HSE has been involved in ergonomics considerations during the course of its everyday work.

1.1 Ergonomics, health and safety

Ergonomics, considered in its widest sense, affects all our work activities, from simple everyday things like sitting and writing at a desk or lifting and carrying a load, through to complicated operations such as controlling a nuclear power station.

The application of ergonomics principles is therefore essential to good occupational health and safety practice for all work activities. While these principles are considered frequently in relation to high hazard industries, known principles are, unfortunately, often ignored at plant or shopfloor level.

We therefore need to get the message across that when workers operate in less-than-optimum conditions they will have to increase their effort to maintain efficiency or even to complete the task. This increased effort may lead to immediate error or overstrain resulting in accident or injury (e.g. back injury from manual handling or lifting too heavy a load, a fall from over-reaching when working from a ladder, operating an incorrect machine control) it may also lead to long-term degradation of performance resulting in impaired efficiency and possibly also to physical and mental ill health, e.g. musculo-skeletal disorders particularly of the upper limbs, some of which may be linked to repetitive tasks.

The 'less-than-optimum conditions' mentioned above may arise from poor machine design, inappropriate workplace layout, poor posture, environmental constraints or from job stress. You are likely to find examples of these problems while walking around any workplace and may also notice situations where employees have made their own adjustments, e.g. work surfaces that have been raised by placing blocks under table legs, the introduction of home-made foot-rests, seats that have been raised by adding cushions, identical control levers that have been classified by the addition of unusual markers or handles.

These 'do-it-yourself' signs often indicate a lack of ergonomic awareness and an underlying problem. Not all employees, or employers, will recognize the problems, nor provide effective remedies; worse still, others will introduce changes which are positively hazardous.

Where particularly the work process is, or is becoming, more complex (for instance, the potentially more hazardous industries such as nuclear power and chemical processing) another branch of ergonomics, that of human reliability or human error analysis, has to be applied. Given a specific situation, particularly a complicated operation under stress, what, then, are the chances of the operator making a mistake? The potential for accidents may be reduced by gaining a better understanding, not only of the work process, but also of the worker.

1.2 Application of ergonomics principles in the workplace

The solution to the problems of industrial injury and ill health is prevention rather than cure. If we accept that applying ergonomics can benefit both the worker and his work, how can we ensure that ergonomics principles are applied at the workplace? This can be done only by convincing industry of the need to apply ergonomics principles, particularly when designing (whether it be machines, plant, workplace layout or the environment) or when drawing-up systems of work. It is only by a thorough examination of work practices that the ergonomics problems (which give rise to accidents or ill health) can be reduced. The chapters in this book address many of the key ergonomics principles, which if applied in a systematic way, would lead to a significant reduction in accidents and ill health. The identification and elimination of problems at the design and planning stage are the measures that are likely to be the most effective and the most economic.

There is no doubt that good design of work systems, tools, equipment and furniture can lessen the likelihood of strain injuries. Anatomical and physical limitations need to be taken into account when new machinery is on the drawing board. Job design is also important to avoid long spells of repetitive or static work.

The proper design of the man–machine interfaces (MMIs) of control rooms and operating procedures, can lessen the chance of human error. The rapid technological change, the introduction of computers and visual display units (VDUs) have had an impact in the office as well as in the factory and it should be recognized that these ergonomics principles can be applied as effectively in the office as elsewhere.

It is perhaps appropriate here to make the point that Section 6 of the Health and Safety at Work etc. Act 1974, obliges designers, manufacturers and suppliers to provide articles and plant that are safe and without risks to health.

Not all problems, however, can be solved at the design stage and, clearly, numerous plants already exist; we must therefore consider the correction of problems from an operational point of view. In many cases, minor modifications to the system of work, to the workplace or to the environment, can be implemented relatively easily to make the work much safer and healthier, e.g. introduction of rest periods, redesign of tool handles, adjustment of the height of a workbench, or redesigning the manual handling process (possibly by replacement with mechanical equipment).

1.3 The Health and Safety Executive's involvement with ergonomics

I would like to mention here just two areas of interest, namely personal protective clothing and equipment (PPE) as relating to protective clothing and the inclusion of ergonomics in the design process for Sizewell B nuclear power station.

1.3.1 Personal protective clothing and equipment

There is a risk in many working environments that health will be impaired from the direct effects of environmental pollutants – dust, chemicals, noise and thermal radiation – where exposure cannot always be reduced or eliminated by enclosure or control. Special clothing or PPE may then be provided for the workers. However, it has often been found that protection is not worn because it is uncomfortable, or interferes with the ability to perform a task effectively.

A particular area of recent concern has been the problem of comfort and fit for head-mounted equipment such as goggles and respirators. In order to address this problem, the HSE commissioned a study of head and facial dimensions. Anthropometric dimensions from nearly 400 individual heads were collected using a computerized measuring device. Measurements were made on both men and women of various ethnic origins.

The HSE's research laboratories have used the data from this survey to generate headform shapes which are representative of the characteristics of particular sections of the working population. A male 'average' headform has been produced from this data using computer-aided design (CAD)/computer-aided manufacture (CAM) techniques. It is intended to use the headform relating to eye protectors, and other headforms representative of other sections of the population, in British Standards for PPE as a means of assessing fit characteristics (Figure 1.1).

Figure 1.1 The 50th percentile Caucasian male headform

While these data provide the information required to improve both the comfort of, and the work efficiency when, wearing such equipment, PPE has often been reported as interfering for other reasons with a worker's ability to do a job effectively. An example of this is the use of safety goggles, particularly in hot environments such as foundries where misting affects the wearer's ability to see (Figure 1.2). A study of misting characteristics in eye protection was undertaken in the HSE's laboratories. Findings from this research resulted in the patenting of goggle designs which would resist misting. Prototypes based on these patents have been field-tested in several industries to assess their effectiveness and acceptability. These field trials showed that, whereas 55% of

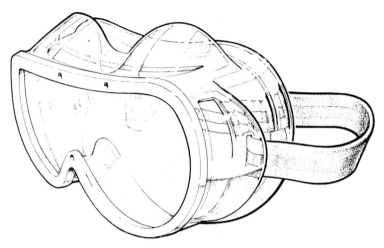

Figure 1.2 The Health and Safety Executive anti-mist goggle

wearers experienced considerable misting problems when wearing their normal eye protection, no one encountered such problems with the anti-mist goggles. Discussions are being conducted with manufacturers to arrange for the production of the goggle under licence.

1.3.2 Application of ergonomics in Sizewell B nuclear power station

Sizewell B is being designed with a target probability for the release of radiation of less than one in a million years. The engineered systems are designed to meet the overall target, and this is demonstrated by a reliability analysis that assumes very high standards of maintenance and operation. However, errors by the operating staff could degrade the designed performance of the engineered systems and this leads us to the question: How do we minimize these errors?

The Nuclear Installations Inspectorate (NII) has identified the following facets of station design and operation as potential sources of human error:

1. Shift staffing levels, and allocation of tasks to the operating staff.
2. Design of the MMIs in the central control room and elsewhere on the plant.
3. Design of maintenance facilities and procedures.

4. Setting and monitoring of operating goals.
5. Staff selection and training.
6. Design of operating procedures and documentation.
7. Management structure and practice.

Particular attention is being given to detailed task analyses of the operational and maintenance actions related to safety functions. The potential for human error is identified where possible; its most severe consequencies are prevented by system design changes for items such as interlocks, and the less severe ones are mitigated by changes presented at the MMI or by modifications to operating procedures and training.

Automatic safety systems are relied on during the 30 min period following a reactor trip or major fault. This is to give the operators an opportunity to diagnose the situation and to plan a recovery strategy – they have time to think whilst not being required to take actions. Studies are being undertaken to consider how to prevent operators taking actions during this period which might degrade the automatic safety systems.

The training of operating staff for fault and emergency conditions is extremely important. The NII is placing considerable emphasis on training for workers at Sizewell B by using a training simulator. The simulator is required, by a condition of the site licence, to be in operation a year before loading of fuel into the reactor.

1.4 Summary

It is clear that the application of ergonomics can make a major contribution in ensuring a safer and healthier workplace. The challenge is to apply existing ergonomics principles in the workplace now and to develop new techniques to overcome existing problems.

We need to see that designers and manufacturers of all types of equipment and plant are better informed and better trained to apply these principles. We must also ensure that, at plant level, managers and supervisors are aware of, and apply, these principles in setting systems of work.

I believe that the challenge is worth meeting – a more careful systematic application of ergonomics principles must inevitably lead to a reduction of the toll and misery of ill health and industrial accidents, and will also improve overall operating efficiency.

Chapter 2

The contribution of ergonomics to present and future industrial safety and health

P. R. Davis

2.1 Introduction

Ergonomics is concerned with the provision of efficient workstations, devices and products that match human capabilities and
needs. Since one of the human needs in industry is safety – both
personal for the worker and environmental for the public at large –
it follows that that which is not safe is not ergonomic. There are
thus areas of common interest between the safety officer and the
industrial ergonomist, as well as areas of separate expertise. If an
operation is safe, the safety officer will be less concerned than the
ergonomist with the productivity of a system and the personal
satisfaction derived from his work by the operator, and the
ergonomist, in his contributions to workplace or product design,
will be less concerned than the safety officer with such factors as
material strengths and conduit efficiencies. The ergonomist
generally has little concern for the inner workings of machines,
provided they do not impinge on the operator or impose hazards
for those who maintain them.

The main contributions to safety deriving from ergonomics
have, thus, come from the application of its extensive researches.
Ergonomics was the child of wartime research into human needs,
capabilities and performance, and for the first half of the existence
of its professional society in the UK more members were devoted
to human research than to its application. This is no longer the
case today, and a recent survey (Ergonomics Society, 1987) shows
that slightly over half the members of the Ergonomics Society, as
in other countries, are now engaged in virtually full-time
ergonomics practice. A large amount of fundamental and applied
research continues to be carried out, nonetheless, the fundamental
part being undertaken mainly in higher educational and defence
establishments, and the applied portion mainly within industry.

2.2 Definitions of areas of safety

One has to define the various areas of safety involved before considering the contribution ergonomics can make. As the ergonomist sees it, there are two types of safety, that for the operator, and that for others. 'Others' may be divided into those in proximity to the operator, and those distant from the operator; 'those distant from the operator' include product consumers and the public at large. In both types – operator and others – there are the effects of time; thus, one has hazards that are immediate or delayed, the delayed subsection including the hazards which arise from time, and those which are a consequence of repetition of activities which impose no hazard if performed once or only a few times.

If one analyses the ergonomics literature one can get a 'feel' for the general size of the contributions made to these areas over the past 20 years, and one thus arrives at Table 2.1.

Table 2.1 Ergonomics research contributions to different areas

Hazards	Operator	Others		
		In proximity	At a distance	
			Consumers	General population
Immediate	+++++	+++	+++++	++
Delayed time only	++	+	++	+
Repetitive	++++	+	++	+

The crosses indicate the relative efforts made by ergonomists

Table 2.1 appears to reflect the priorities given to safety by many other groups working in industry. It does not, of course, reflect the perception of safety requirements of the public at large, but does show the scope of present ergonomic knowledge of value in the safety field. It also indicates that hazards to the general population, while being of high concern, have received the least attention by researchers.

2.3 Definition of ergonomics research

The research reviewed here has been undertaken by members of ergonomics and human factors societies over the past 40 years.

There were originally very few people in these various societies who considered themselves to be ergonomists, but most had a deep interest in the subject. Thus, the main initial contributions came from psychologists, physiologists, anatomists and defence and industrial medical personnel. The passage of time has seen a regularly increasing number of qualified ergonomists, and the proportion of research contributions from recognized ergonomists has increased in parallel with this. Since some safety personnel are also members of an ergonomics society, their activities may overlap with those engaged in standard ergonomics activity. With such an enormously wide scope, it is hoped that those whose work is not referred to below will not feel too badly ignored; some limitation has had to be placed on the selection of topics.

2.4 Ergonomics safety contributions

That which is unsafe is not ergonomic. Many accidents and injuries at work arise when the work arrangements are not fitted to the workers' capabilities or interests. Misfitting may be physical or psychological. Physical misfitting can cause such things as overstrain of the musculoskeletal and cardiovascular systems, and spatial misfitting can cause such things as postural stresses, over-reaching or bumping contacts (Figure 2.1). Psychological misfitting is less easy to determine, but may have much wider effects than simple physical causes. Sensory misfits may directly affect vision, hearing, olfaction or any aspect of peripheral sensation, or more insidiously affect the central processing of sensory information. The latter can then disturb cognition, and may affect reaction times and cause wrong decisions to be taken. Equally, such matters as ill-chosen shiftwork times and rest pauses can lead to disturbed thinking and, thus, to serious errors, as can lack of appreciation of the effects of biological rhythms. Knowledge of these factors is part of the ergonomist's armoury for designing satisfactory and, hence, safe workplaces, and this research has contributed much to safety considerations. These many contributions are considered above in relation to Table 2.1.

2.4.1 Immediate hazards

The ergonomist, while having concern for matters contributing to immediate dangers such as the coefficient of friction at the shoe–sole interface and the lack of splintered surfaces and sharp edges, has been more concerned with ensuring that the workplace

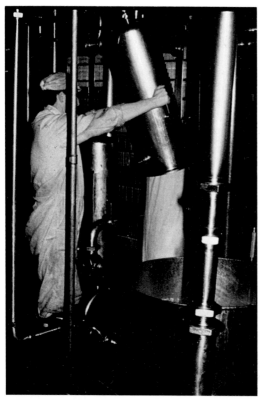

Figure 2.1 Postural and spatial stress: when removing or replacing the cover to this filter, the worker is required to handle the load at arm's length; the positioning of low level pipework prevents him from moving closer

matches the size and abilities of the worker, and that the worker is able to control tools or machinery without posing immediate personal or social hazards.

There have thus been major anthropometric studies of males and females of different ages in many working populations, some of these culminating in the production of guidance mannekins, or definitions of spatial optima and their variances for the whole, and different parts, of the human body, in a large variety of occupations (see, for example, Pheasant, 1986). Studies of the space requirements for normal and laden locomotion have also led to considerations such as tread heights and depths for stairways, the slopes of ramps and their widths in relation to task needs. In the field of physical work, tissue, muscle, limb and whole body strengths have been determined – again for a number of different populations – and a reasonable picture of the spread of physical capacities is now available (see, for example, Chaffin and Andersson, 1984). Physiologists have investigated the cardiovascular abilities of working people, and have measured the energy demands of many tasks and occupations. However, it is clear that much more knowledge regarding females is required, partly because of past lack of interest, and partly because in Western societies their physical capacities appear to be increasing faster than those of males.

The effects of posture and load size, shape and weight have all been studied widely, and the optimal sizes and locations of handles for various loads have been determined. The physical effort required for many tasks has been measured in this regard, and the effects on these of altering tool or lever dimensions, required force and shape have been recorded. One factor perhaps worth highlighting in relation to immediate hazard is the interplay between the magnitude of effort and the perception of the adequacy of its control.

Much attention has been paid to visual abilities. The needs for optimal acuity for different types of task have been defined, and the effects on these of factors such as luminance, glare, flicker and colour are well documented. The optimal sizes of print for different purposes has been established, and the legibility of a range of fonts compared. The factors involved in clear visual differentiation of control functions have been determined.

Acoustical studies have included the effects of noise on speech recognition, allowing prediction of error rates at different levels of interference, and the effects of different levels of noise on cognition, decision making, learning processes, and such matters as calculation and hand–eye co-ordination. Criteria for the efficiency of protective devices have been established, and it is now possible to prescribe optimal sound levels for given tasks.

Much of the initial research in ergonomics was concerned with controls and their associated feedback devices. Indeed, there was a time when the general public thought that ergonomics was 'knobs and dials', but fortunately that partial misconception has passed. Nonetheless, control systems continue to be an important part of ergonomics and research continues. The original observations included studies of dial sizes, types and positions, the best sizes, colours and uses of knobs and levers, and the optimal placements of these in relation to frequency and/or urgency of use. Such observations led on to the realization that every control situation had to be thought of as a 'system' rather than as a series of isolated objects, and this involved the development of studies into hand–eye co-ordination, reaction times, and the whole area of decision making. Many case studies illustrated the possible immediate adverse effects of bad provision of control devices, some of these being of catastrophic proportions. The increasing complexity of industrial systems had to be met by studies of human capacities to absorb, remember, digest and act safely on multiple sources of information.

Inherent in all these matters has been the question of fatigue and the effects of stress on functional integrity and safety. Many investigations have been reported into the effects of physical and mental fatigue on all types of performance, and some of these have been associated with parallel studies of the interplay between direct fatiguing factors and environmental conditions. Various measures of fatigue have been postulated and assessed, and although none of these has been of universal application, the battery of tests now available allows fatigue measurement in most circumstances. Thus, accidents arising from this cause can generally be ruled out at the planning stage.

In addition to these contributions to the safety of industrial workplaces, there have been many studies of work systems, including analyses of individual factory departments, factories as a whole, groups of factories and entire industries (Figure 2.2). These

Figure 2.2 Gathering ergonomic data in a cable manufacturing plant for use in future workplace design

have revealed a need for such wider studies, in that in many instances it has been found that recognition of an immediate hazard at one workstation has caused an alteration in the work procedure, resulting in a new hazard in the next stage of the operation. Often these effects are complex, but one simple example may illustrate the importance of wide system studies. One common finding has been that such a chain reaction has affected the transportation, wholesale and retail parts of a system. In one case, palletization by a producer increased productivity and lessened the risk of injury in the factory, but it increased the risk of immediate injury for the wholesaler's staff, and matching provision of forklift trucks and fitted storage in the warehouse had to be provided to improve immediate safety there. However, deliveries to individual retail outlets then became more dangerous, as many smaller concerns were not, and could not, be equipped with forklift trucks and hoists, and pallet handling and unpacking in these circumstances increased the risk of immediate injury for the shop staff.

One thus can claim that ergonomics has contributed much to the prevention of immediate hazards in working situations. A workplace based on ergonomic design will be well ventilated and lit, the noise level will not intrude on clear communication and thought processes, the space available will fit the needs of the operator for all the activities associated with the job, the flow of information from the machinery will be clear and logical, the controls will be placed in the best positions, the forces to be exerted will be well within human capacities and the input materials will be at the correct locations for safe handling, as will be those for the output. Rates and times of work will not lead to unwanted fatigue. The corridors, doorways, slopes and ramps will be adequate, not only for normal work movements but also for emergency procedures. Any protective clothing, respirators and similar devices will fit properly, and safety devices such as alarms and extinguishers will be simple and quick to use (many require much improved design to this end). Given such provision, the level of immediate hazard is very low and is probably the best that can be achieved.

There is, however, one other factor which has also to be considered in many occupations, and it is one which has received much attention, and may clash with the provisions outlined above. This concerns the level of vigilance. Many studies of control and inspection tasks have been reported, and the factors underlying the proper levels are fairly clear. Often failure arises from the workplace being too complex, but it has also been found in some cases that a fully fitted workplace may be too soporific for safety. One early example in Germany was a factory power station in which far too many human errors had occurred because the dials were difficult to read, the controls were badly designed, and both dials and controls were so placed that they bore little relationship to each other. An ergonomic redesign was carried out, and initially the error rate dropped dramatically. Unfortunately, it then started to rise again and the ergonomics team was recalled. They found that the effort required of the operators was now minimal, and the place was so quiet and the seats so comfortable that they were going to sleep and missing important changes. The answer chosen was to place the seats at a distance from the consoles, so that the operators could get the necessary physical rests, but had to move to and fro and stay awake as well. There have been many studies of the factors underlying good levels of vigilance, and the maintenance of such things as adequate motivation, and any safe workstation design will include such considerations.

In Table 2.1, the immediate hazards to others are considered. Those in the proximity are generally covered by the researches outlined above, and need no special reference here. At a distance, the consumer has received much attention, and the importance of ergonomic contributions to product design has been reflected in the setting up of the Institute for Consumer Ergonomics at the University of Technology at Loughborough. The aims of all ergonomic consumer research include a very large consideration of product safety, and such work has taken into account the unfortunate fact that many consumer products may be misused by the general public. Since the general public is again to a large extent covered by the studies of consumers, and of those others presented above, less direct attention has been paid to their needs, but ergonomists have made important contributions with their studies of such things as crowd behaviour, public behaviour at railway stations, in marine catastrophes and similar situations. Appraisals of designs for cars, lorries, buses and trains with regard to entry, internal movement and seating have improved safety levels for the public. The speeds of elevators, moving walkways and open step-on lifts have been observed, and acceptable velocities for different populations have been established. Again, the work on information systems has improved public communications, leading to many improvements in personal safety when travelling. In this same area, ergonomists have been deeply involved in the human factor side of airport design and of passenger aircraft, and have had much responsibility for human aspects of air traffic control. There have been important studies of pedestrian behaviour of both adults and children, particularly in relation to road safety. Thus, ergonomists' contributions to public safety have been considerable.

2.4.2 Delayed hazards

Delayed hazards include such adverse effects as accumulated toxins, chronic radiation overdosage, cumulative skeletal pathology and long-term visual strains. The contributions of ergonomics to some of these problems have been small, but there have been major advances in the control of others. Those resulting from adverse conditions only as a result of time have led some ergonomists into the difficult field of ageing, since some effects are barely distinguishable from normal ageing processes.

One example has been the use of epidemiological studies in parallel with ergonomic analyses of workplaces which have led to recommendations regarding future designs. One instance here is the redesign of an oil-can filling machine, in such a way that the operators handling the cans could not become contaminated and the risk of chronic skin lesions was greatly reduced. Some of the bigger contributions have been in the area of atomic energy, where major ergonomic appraisals of power stations, including control systems and individual workstations, have improved work efficiency and considerably decreased acute and cumulative radiation hazard. It is interesting in this regard to note that full-time ergonomic input has been arranged into the planning of operational safety in new British nuclear power stations, partly as a result of intercession by the British Ergonomics Society.

Another major contribution has been the research into cumulative physical trauma. Epidemiology has shown that some physical stresses may be well within safe capacity when they occur only once, but that if repeated over periods of hours, days or years may have serious effects. These include such injuries as vibration white finger, tenosynovitis, cervical spondylosis, and a significantly large proportion of cases of back strain. Much effort in recent years has been devoted to establishing force limit tables whose application should help to obviate such injuries. The guidance limits so far have been concerned with short- and medium-term effects; lifetime exposure has yet to be studied fully. However, short- and medium-term guidance levels for some vibrations are fairly well-established, as are those for the prevention of cumulative back injuries (Materials Handling Research Unit, 1980) and are becoming incorporated in guidance documents across the world; those for other disorders are still being investigated, but it seems likely that our knowledge of limits for repeated limb and head stresses will soon result in similar safety indicators. The methods used in back research include psychophysical studies, in which the subjects judge for themselves what forces they would be prepared to apply: a comparison of these results with those from other methods shows that, for some actions, subjects will apply repeated forces, which an investigation by Nicholson (1986) suggests may be hazardous. This makes it clear that external measurement of physical work is necessary for full safety.

There is a growing realization, also in relation to delayed skeletal disorders, that prolonged maintenance of posture can lead to cumulative injuries. While there have been many studies of

Figure 2.3 Prolonged poor posture leads to muscular discomfort and poor work output

sitting posture and support design, many of these have been concerned with short-term effects. Some of the epidemiology referred to above has shown that workstations which demand long continued maintenance of a rigid posture of any kind can cause immediate or short-delay discomfort, and also have an associated high rate of delayed injury (Figure 2.3). Data are emerging which can be used as indicators of safe tolerance times for such situations.

2.5 The future

In relation to industrial safety the future of ergonomics is likely to consist of two types of effort. The first is concerned with safety systems and devices. There is a considerable leeway to be made up with existing equipment. For instance, there are indications that the general public have a stereotyped image of a fire extinguisher which differs widely from reality in many cases, and the same would seem to hold for some who have responsibility for fires at their place of work. This is an ergonomic problem, and will need to be answered in ergonomic terms. Future provision of safety equipment for, as yet, unheard of hazards will also require consideration in an ergonomic context. One can only guess at some of these: the biologists using genetic engineering begin to pose new problems as operations which, up to now, have been confined to the laboratory are enlarged into production processes; the detection of, and protection from, escapes of micro-organisms and their products, an area which will need much further research for both safety personnel and ergonomists, and may include a need for tolerance studies for environments which may be weakly harmful to man and either highly toxic or highly beneficial for these organisms or substances. While genetic engineering is one likely example, there will be many others, but the only way to deal with the hazards will be for managers to keep their safety and ergonomics staffs fully informed from the beginning, and for these practitioners to keep themselves well informed as to the industrial application of scientific advances.

The second type of effort must be concerned with the improvement in utility and safety of existing and future products. This is already an established need in the UK where growing dependence on income from exports will place further stress on this requirement. The safety engineer and ergonomist have played only secondary roles in too many of our manufacturing industries with the result that our exports have either been unfitted to the safety requirements of the receiving countries, or have been of less utility than those of our rivals. In more enlightened companies, both are consulted at an early stage in development, and the resulting product sells well in the overseas market. Thus, both types of practitioner must, and I am sure will, make themselves conversant with international safety requirements and with the anthropometry and behavioural peculiarities of those receiving and using our goods.

In parallel with these activities, one can see that there is a burgeoning amount of development of manufacturing machinery and consumer goods for virtually exclusive sale to what are presently known as the under-developed countries. Many of these countries are only under-developed in the sense that they need good equipment which is less sophisticated and more robust than that required by more advanced workforces, which fits their body sizes and is tailor-made for their needs. If we are to take advantage of this potentially huge market, then all of us – manufacturer, ergonomist and all safety personnel – will have to know more about these countries and the needs and activities of their populations. While most manufacturers are prepared to modify existing products to partly meet local needs, it is only the producer who designs a product fully satisfying for the customer who will make a large contribution in this field. The question of the environment is inherent in some of these areas, and here the ergonomic literature has many papers concerned with the effects of tropical and Arctic environments on human behaviour. These make it clear that studies of new man–machine interfaces (MMIs) in such exotic environments will be all-important in obtaining fully satisfactory and safe designs.

The reader may have been surprised that, so far, little attention has been paid here to the question of computers and their uses. Much of what has been outlined in the sections on immediate and delayed hazards applies to, and has been applied to, computer design, and to other keyboard-based appliances. There is a large group of ergonomists within the British Ergonomics Society with interest and expertise in the use of these machines, and there is considerable international activity as well, of course. But computers are in a transitional stage, and are more for the future than the present. The ergonomics of existing domestic and small business machines is unsatisfactory, in that the keyboards generally are sub-optimal; the instruction manuals are often incomprehensible, and give no guidance as to proper installation, seating needs and lighting. It has been said that there are more abandoned machines of this type than those of all types actually in use. Most users complain of eye strain and postural discomfort, and we have yet to see whether this will lead to extensive permanent damage. No one can deny that these machines will extend their deep influence on all aspects of society, and the existing research effort will clearly increase rapidly. Microprocessors are used to improve various performances in vehicles and domestic machinery and where they remove the need for human judgement they may be very welcome; however, in some situations they require human input, and then they need careful matching to the user's abilities. One of the reasons for their inclusion in a design is that of speed of response, and in this it is sometimes found that microprocessor + man is slower than man alone, as the information flow from the machine to the man, and the speed of human comprehension, are together slower than if the ordinary controls had not been replaced. Selection of process must match human capacity if efficiency and safety are to be achieved.

Another area, connected with the foregoing, which will need much further work, is that of robotization. There is a tendency by engineers and managers to think that, because a process has been robotized, it no longer needs consideration in relation to human factors: nothing could be further from the truth. All robotized systems have to be made, installed, maintained and serviced by human beings, and the accident and injury rate per man-hour has not fallen with these machines. The safety and efficiency of the installer and, particularly, the maintenance engineer, thus needs much more consideration than it receives at present. Indeed, much conventional machinery poses unnecessary hazards and difficulties in maintenance, and could have been much improved by better design.

One may be taken to task for not including ergonomic aspects of training in the main body of this chapter, but many papers and publications relating to training have appeared. One of the dicta of ergonomics is that training in safety procedures is only necessary where there are ergonomic deficiencies in the workstation, and, one could say of some managements that it has been cheaper to print posters than to remove an easily obviated hazard. Some safety training is necessary now and will remain so, but there is a growing awareness of the need for safer design at all stages of manufacture, and in the supply and service industries, and ergonomics will continue to play an important part in this.

Lastly, perhaps one should look further into the future and consider some of the formidable social changes that are happening, and the related areas of safety which will require the attention of ergonomists and safety practitioners. The rising rate of population growth is leading inexorably to increased numbers of metropolitan areas with ever-increasing population densities. Even if satisfactory answers are found to the dreadful problems of exploitation and pollution posed by this change, the problems of highly crowded living will remain and intensify. All safety workers are only too well aware of the dangers posed by overcrowding in homes, factories and movement areas, but the studies on these subjects have been concerned with population densities far below those now extant in some parts of the world and which will inevitably increase in many countries. There appear to be no comprehensive studies of the safety, health and ergonomic aspects of high-density living aimed at producing acceptable tolerances for space limitations, degrees of sophistication, personal contact frequencies, disease limitation and so on in what we would today have to consider as severe overcrowding. And yet it is an area which will continue to force itself upon us, and our researches may then help to solve Man's present greatest problem, Man himself.

2.6 Conclusion

Ergonomics has made, and will continue to make, many important contributions to industrial safety. There is room for improvement in many existing work situations, yet knowledge of ergonomics can supplement that of the safety professions in realizing that improvement. There is much active ergonomic research into areas of ignorance, and this seems likely to continue. It is also clear that close co-operation between the professional societies of ergonomists and safety workers will be required to meet the challenges of the new industrial processes and ways of working that seem likely to develop in the near and later future. For proper safety, early inputs of such expertise will be required during developments.

While the engineer and scientist can achieve near miracles in their use of the Earth's resources, they will only be of safe benefit to humankind if human needs are kept to the forefront in all endeavours.

References

CHAFFIN, D. B. and ANDERSSON, G. B. J. (1984) *Occupational biomechanics*. Wiley, Chichester

ERGONOMICS SOCIETY (1987) *The Ergonomist,* 205 and 206

MATERIALS HANDLING RESEARCH UNIT (1980) *Force limits in manual work.* Commission of the European Communities and Materials Handling Research Unit, University of Surrey, Guildford

NICHOLSON, A. S. (1986) 'Manual handling limits – a comparative study'. In W. Karwowski (ed.) *Trends in ergonomics/human factors,* vol. 3. Elsevier, Amsterdam

PHEASANT, S. T. (1986) *Bodyspace: anthropometry, ergonomics and design.* Taylor and Francis, London

Chapter 3

A European perspective on ergonomics and health and safety

J. G. Fox

In determining the content of this chapter, I first had to decide on what would constitute 'Europe'. Was it to be the total geographical land mass and its offshore islands, or Western Europe as identified by the European Economic Community (EEC)? In taking the EEC should we pursue developments in the Community as an institution: or would there be benefit in looking at individual member states or other institutions within the Community?

For reasons of clarity I shall restrict this text to the Community, its member states and associated institutions.

Secondly, what were to be the criteria for the 'perspective'. I, more easily, arrived at a single criterion. Several years ago, I was honoured to open the Ergonomics Society's annual conference and I chose as my theme 'The establishment of ergonomics within the Establishment'. Today I would like to use this theme as my yardstick here in presenting a 'European perspective'. Whatever ramifications of ergonomics individuals or institutions have found it convenient, or profitable, to pursue, ergonomics, by definition, is about 'people at work'. The only measure of its importance and its success lies in the degree to which its concepts and practices have entered the legal and socio-political fabric of our work systems, be they manufacturing, service or consumer-based.

The EEC has made steady, if unspectacular, progress over the last 10 years in expressing the political will for the development of ergonomics within the context of health and safety at work. In parallel, within the European Coal and Steel Community (ECSC), ergonomics has developed to become a major force in health and safety within these two industries. The difference in speed of action is explained by the fact that while the Treaty of Paris, which established the ECSC, made a commitment to the health and safety of the workforce of the steel and coal industries, the Treaty of Rome, which established the EEC, had no similar articles with respect to the workforce in general. Nevertheless, over the years the Council and the Economic Committee of the EEC have recorded their wish to see ergonomics action meet the needs of health and safety at work, particularly with the advent of the new technologies. For example:

1. The Council, in its conclusions on 'Technological change and social adjustment' on 7 June 1984 requested the Commission to initiate work to encourage the systematic acknowledgement of their ergonomics implications with a view to improving working conditions, starting at the stage of conception and planning of technological innovations.

2. On the 'Medium-term social action programme', the Council concluded on 22 June 1984 that: 'Arrangements should in particular be sought for taking systematic account of the ergonomics implications of technological innovations as from the *stage* of the study and design of equipment.'

3. The Standing Committee on Employment, in its paper on 'New technology and employment' published in 1985, noted that: 'The work that has been done in the field of ergonomics can help improve working conditions.'

These expressions of political will have today become more than pious hopes.

The 'Single Act' which was ratified in 1987 made a number of modifications to the Treaty of Rome. Much was made by the media of the threat it might pose to national sovereignty. Article 21, however, passed through almost unnoticed. This article of the 'Single Act' created a new Article 118A to the Treaty of Rome, which provides that: 'Member states shall pay particular attention to encouraging improvements, especially in the working environment, as regards health and safety of workers, and shall set as their objectives the harmonization of conditions in this area, while maintaining the improvements made.'

Following in the wake of this upgrading of industrial health and safety in the Community policies, the Commission will almost certainly have introduced a 'Third programme of health and safety at the workplace' by the beginning of 1988. This programme will, perforce, have an authority which its predecessors did not. Its principle approach will be appropriate directives and guidelines for the member states. Its ergonomics content will address itself to information technology, process control and biomechanical damage risk at the workplace. More general directives will also have an ergonomics content. In this context, are directives, well advanced in their preparation' on 'the essential safety requirements relating to the design and construction of machines' and 'the organization of safety at work'. While I would not wax over-euphoric on these advances for ergonomics (competance and numbers in the Commission staffing are far from adequate to sustain them) it gives cause for some satisfaction to have heads of state and senior ministers cognizant with ergonomics.

These developments in ergonomics within the Commission have their counterparts in other institutions having relevance for Community member states, e.g. ILO (International Labour Organization), WHO (World Health Organization) and international trade union federations such as FIET (International Federation of Commercial Clerical, Professional and Technical Employees), the white collar workers' federation.

The most recent organization to join this group is the Committee for European Normalization/Standardization (CEN). The CEN has formed a new technical committee, TC/114 Ergonomics. The implications of this event are very significant and too detailed to pursue here; I would simply recall for you that, unlike International Organization for Standardization (ISO) standards, CEN standards are mandatory.

Major modifications to Community policy do not occur gratuitously or by goodwill. They emerge from pressure by member states. The changes relevant to our present discussion come from pressure by states where already there exists a strong commitment to the social and economic necessity for safety, health and ergonomics in the industrial society of the country: states where, perforce, the social affairs lobby is strong and/or the trade unions pay more than lip-service to the protection of their workers' health and safety. In Holland, for example, the 'Arbo-Wet' (the law passed in 1980 concerning working conditions) holds the engineering designer responsible for hazardous situations which are the consequences of his design.

In France, there is an absence of detailed legislation. This has not, however, prevented the *Force ouvrière* and other trade unions in several sectors of French industry and commerce negotiating collective agreements with respect to ergonomics factors in their work situations.

West Germany probably has the most comprehensive systems of regulation of health and safety with a strong ergonomics content. The principal regulatory mechanism is probably not the state but the *Berufsgenossenschaften* (self-regulating groups of employers called trade co-operative 'associations'), whose regulations have the force of law and attract penalties for non-compliance both by employers and workers. These regulations are viewed in the light of the state of the art as determined by the very important Deutsches Institut für Normung (DIN) standards which virtually cover all current aspects of ergonomics. As an additional check, the Constitution of Enterprises Act of 1972, among other things, gives the right to employees' representatives to evaluate and (on justification) to veto the introduction of any equipment or system which does not meet ergonomics standards. Much meat has been put on these infrastructures by the implementation by the Federal German Parliament of programmes of 'humanization of work'. The first during the early part of this decade had an annual budget of DM80 million and a second was approved in 1985. The programmes themselves deserve to be held in awe but more important is the follow-up. Three specialist institutes have been set-up to evaluate the results of the first programme and to ensure the necessary technology transfer to the industries.

These national and international developments have provoked direct responses to specific ergonomics requirements in ensuring health and safety within various sectors of industry and commerce where it did not exist before. A case in point has been the ergonomic requirements of high-technology office systems. The response of the international information technology industry was in no small way influenced by the position and action taken by West German commercial users and trade unions and backed by their legal framework. Ergonomics is now a key feature in product design considerations of information technology hardware, and software as a result. From a recent study I commissioned, the rather surprising fact emerged that on the basis of personnel, approximately 65% of the claimed expertise in the Community in the field of ergonomics and information technology now lies with the manufacturers. Among the available volumes of ergonomics guidelines for visual display units (VDUs) there is one from the European Computer Manufacturers' Association and yet another from the European Association of Manufacturers of Business Machines and Data Processing Equipment.

Earlier I mentioned the development of ergonomics as a major force in health and safety within the ECSC. I would like to enlarge on this comment and draw some conclusions from the experience of some 13 years directing three successive ergonomics programmes for the ECSC.

Ergonomics is not unknown to the ECSC. Its first programmes in health and safety were begun nearly 30 years ago and included 'human factors'. Over the last 10 years, £42 million have been committed by the ECSC to improve safety and health in their industries by way of a programme of ergonomics action – a programme committed directly to achieving ergonomic change at the workplace.

It must be more than surprising that ergonomics showed much development or, indeed, survived as a significant factor in the coal and steel industries against an economic recession that reduced job opportunities in Western Europe to a total unknown in 50 years. That ergonomics showed growth was due in no small part to its 'action approach', which demonstrated that the quality of working life, health and safety are in no way in conflict with the creation of wealth: that, indeed, wealth will the more readily be produced in humane conditions.

The European steel industry has undergone major restructuring during the 1980s with the aid of ergonomics, which started from 'the ground up'. Visits to Germany, Italy, Luxembourg and Holland, for example, will show how new technology in steelmaking was matched with its ergonomics requirements. The prospects for the immediate future can only be described as professionally exciting. A further £17 million will be committed in the current programme to ergonomics, and projects in the steel industry include:

1. The establishment of an expert system for computer-aided design (CAD) of ergonomics of steel processes.
2. The ergonomics parameters for a computer-assisted maintenance control system which includes speech recognition.

Coming from behind, the mining industries of the ECSC have, comparatively, made even greater advances in ergonomics in having to cope not only with new technologies but also with many traditional problems.

Thus, projects such as: (1) 'the ergonomics of software for the automated guidance of shearers; have run alongside (2) 'the ergonomics of cleaning railcars'.

The output from these programmes has, in the main, been ergonomically designed systems and hardware. There is little report of it in the 'ergonomics scientific literature'; but then, it is remarkable how difficult it is to achieve the criteria given in this literature with a proven effective application of ergonomics.

The success of this decade of work in the ECSC has rested on a number of factors, but for the present I would highlight one: a real understanding by all concerned of the nature of the ergonomics contribution to the socio-economic fabric of these two industries. This is a contribution where:

1. Pragmatism is as valued as scientific rigour.
2. Problem-solving replaces an open-ended preoccupation with methods and concepts and has its own intellectual respectability.

3. Solutions are functional rather than theoretical.
4. The social partners replace polite interest by commitment and intelligent appreciation.

It is clear that in many quarters in the EEC there is a consensus that ergonomics plays an important role in our industrial society. I have omitted any reference to the UK because I feel a comparative exercise would be useful. But I am content to leave you to draw the comparisons.

References

EEC (1984) 'Conclusions of the Council concerning a Community medium term social action programme'. *Official Journal of the European Communities*, 34/C, 175/01

EEC (1984) 'Conclusions of the Council on technological change and social adjustment'. *Official Journal of the European Communities*, 34/C, 184/01

EEC (1986) *The Single European Act and Final Act.* Office of Publications EEC, Luxembourg

Part 2

The operator in the system

Chapter 4

Assessment and prediction of human reliability

D. E. Embrey

4.1 Introduction

Everybody has heard about human error, yet people in general are unfamiliar with the concept of human reliability. Humans are actually remarkably reliable, and correct actions are far more common than erroneous ones. However, error is newsworthy; the consequences of error can often be far-reaching. There have been a number of disasters in recent years where human error has been implicated as the major cause: Three Mile Island, the Chernobyl incident, the *Challenger* space shuttle disaster, and the sinking of the *Herald of Free Enterprise*, can all be seen to be failures or errors by individuals or organizations and have justifiably received high media coverage. The same cannot be said of human reliability. How often do we hear or read of pilots who have successfully completed aircraft manoeuvres, or the numerous vehicle drivers who safely arrive at their journey's end?

This chapter will introduce the theory of human error. It is necessary to understand why people make errors in order to gain a basic understanding of how to improve human reliability. It will then be possible to consider the type of human reliability techniques that are available and the likely future developments in this area.

4.2 The nature of human reliability

The study of reliability is concerned with reducing the likelihood of errors by predicting what could go wrong when the human interacts with a work system. This is achieved by applying systematically the principles of psychology, ergonomics and organizational design, together with a basic understanding of the operation of the work system. The essence of the discipline is the prediction and mitigation of error with the objective of optimizing safety, reliability and productivity.

Human reliability techniques have traditionally been applied in high-technology systems areas such as aerospace, nuclear power and the offshore oil and chemical processing industries, but these techniques are applicable in all areas where human error can compromise production or safety. The study of human reliability can broadly be subdivided into two main streams: (1) the qualitative aspect, which is to do with defining what sort of human errors will occur and how to mitigate them; and (2) the quantitative aspect, which is the prediction of the probability of error occurrence. I will return to these two aspects later in the chapter. It is also worth pointing out that human reliability techniques can be used predictively, to anticipate errors and therefore permit preventative strategies to be developed, but also retrospectively, for analysing why major failures or accidents occurred. Before considering techniques which can be used to study human reliability, we need to consider the nature of human error.

4.3 What is human error?

It is a widely held view among researchers in human reliability that all of us have built-in tendencies for error. We are required to live and work in a complex world which taxes the human cognitive system. Our cognition and the actions we take as the result of our decision making are correct on most occasions, but on others we can be led into making mistakes. Even so, a very high proportion of errors that we make are recovered – in other words, we are able to take a course of action, once an error has been made, which overcomes the failure. An error only becomes manifest in an 'unforgiving situation', which ignores human strengths and limitations. For example, an unforgiving situation may be one in which the operator is not provided with adequate feedback information from the system, or which does not take into account the way that people process information or which provides only one pathway to achieving a goal. Since humans are goal-orientated rather than means-orientated and prefer to have a choice of pathways towards an objective (and also because we are extremely innovative in such matters) other pathways will be sought if only one is provided. This can create major problems if the system has not been designed to accommodate such behaviour. We can still learn much about the root causes of errors from such occurrences, even if an error is recovered and, therefore, has no observable consequences on the system with which an individual is interacting.

4.3.1 The underlying causes of errors

We have to consider the nature of the cognitive processes involved in performing tasks in order to understand why we make errors. In conceiving and carrying out an action, three stages are involved: (1) planning; (2) storage; and (3) execution. Planning is the process by which we identify a goal and decide upon the means of achieving it. Such plans are not often acted upon immediately and so the second stage – storage – is required. This can be for variable lengths of time and allows other factors to intervene between the formulation of plans and their execution. The final stage – execution – is the process of implementing the plan. An error can arise anywhere within these stages and can therefore be defined as the failure of a planned action in achieving an intended goal because the action did not proceed as intended or because the planning was wrong. If the planning was incorrect this is referred to as a 'mistake'. If the action did not proceed as planned, this may have arisen as a consequence of either a memory failure – in which case it is referred to as a 'lapse' – or because of a lack of attention – which is called a 'slip' (Table 4.1). Errors therefore fall within two types: either those which occur at the level of intention ('mistakes') and those which occur during subsequent stages ('lapses' and 'slips'). These two categories of error type require different mitigating techniques.

Table 4.1 The primary error types. (After Reason, 1987)

Cognitive stage	Primary error type
Planning	Mistakes
Storage	Lapses
Execution	Slips

4.3.2 Information processing

In order to understand error causation we need to consider how we process information which we receive while performing a task. Analysis of this processing is at the heart of the prediction of human error and, therefore, its mitigation. Rasmussen (1982, 1983) distinguishes between three generic types of information processing: (1) skill; (2) rule; and (3) knowledge-based processing (Figure 4.1).

Skill-based processing arises from experience with a specific situation where particular configurations of frequently occurring

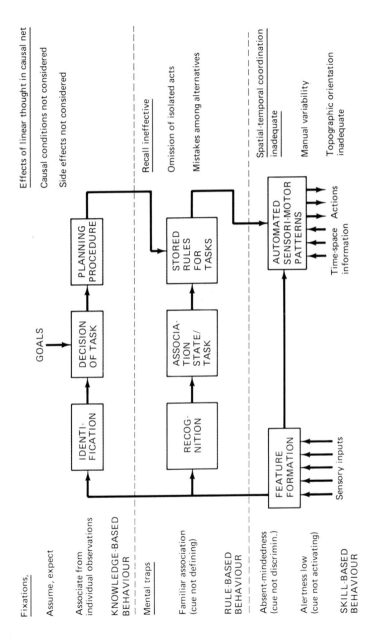

Figure 4.1 A model of human data processes and typical malfunctions. (From Rasmussen, 1982)

symptoms become associated with particular causes. This mode of operation often occurs in situations involving mainly simple physical operations of controls, or where equipment is manipulated from one position to another. No conscious planning or monitoring is required to execute the actions. Errors arise in this mode if an operator produces a stereotyped response to a pattern of indications which is superficially similar to the familiar pattern but which actually arises from a different cause. If, as in certain situations, the operator does not recognize a pattern of indications, this usually prompts a process of scanning and information collecting from various sources. This information is used as input to an explicit or implicit rule of the form:

<IF> Situation X
<THEN> Do Y

This leads to the selection of an appropriate action which will be executed to achieve the desired result – it is referred to as rule-based processing, since this mode of behaviour involves rule-following. A simple example would be: if your car fails to start you may use the following diagnostic rule: if starter motor is OK, but lights dim when operating the ignition key, the cause is a flat battery. Appropriate action can then be taken. Rule-based processing is less common than skill-based behaviour but still occurs very frequently. If the operator cannot refer to an existing procedure or rule of thumb, even after the information-gathering exercise, then he or she enters the third stage of processing – knowledge-based behaviour. In this mode, with no past experience to fall back on, the operator has to utilize his or her overall technical understanding of the situation and formulate new hypotheses from first principles. As a general rule, to achieve highly reliable systems, people should not be expected to operate in the knowledge-based mode when under conditions of stress or pressure of time.

The type of processing which is required for a particular task depends on the individual's level of training and experience. Thus, for beginners, all operations involve a high proportion of knowledge-based processing because they are new to the situation. With practice, most operators move into the rule-based domain, and with some very frequently encountered operations they are able to become highly automated and thus exhibit skill-based behaviour. It should be emphasized that these types of behaviour occur in practice as a series of tightly coupled, iterative steps, with frequent changes between different levels of processing.

Some common situations will help to illustrate these points. Take the case of changing the oil filter on a car, if you have carried out this task many times on the same car, you will be able to operate primarily in the skill-based mode but may need to revert to the rule-based mode periodically by referring to the handbook. To take another example, consider driving home from the office. You may have done this many times before and do not have to think about the directions you are taking; indeed, your mind is probably focused on some completely different matter. You are operating in a skill-based mode. Every now and again you check on your progress by identifying familiar landmarks and if something happens which diverts you from your normal behaviour then you may revert to the rule-based mode. If, for example, the road is blocked by an accident, then you have to consider your options and apply a stored rule. If you know an alternative route home, then, once again, you can revert to a skill-based mode. Otherwise, you may have to go back to first principles and plan out a new route by referring to the map. This requires operating in the knowledge-based mode. Once you have sorted out the new route then you can revert to applying rules (following the map) until such time that you reach familiar territory and you can slip again into automatic behaviour.

Within these processing modes, various errors can be made. In the skill-based, or highly practised situation, it is possible to

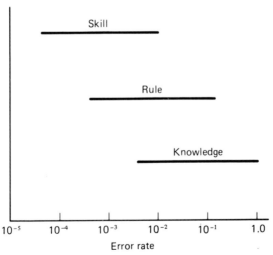

Figure 4.2 Error rate ranges associated with human behaviour. (From Rasmussen, 1980)

exhibit what are called 'strong but wrong' action sequences. This could occur, for example, if you move to a new house where part of the journey home is common to the route you used to take. For the first couple of days after the move there is a danger that you will find yourself outside your old front door unless you make a conscious effort to follow the partly unfamiliar route. This is particularly likely if you have been distracted at some stage to the extent that your actions have been captured by this strong action sequence. Analogous errors can occur in rule-based tasks where strong (i.e. successful and frequently applied) rules can replace the correct rules. In knowledge-based processing, error forms are much less predictable. Some knowledge-based errors arise because the individual only has a very narrow 'keyhole' vision of what may be a complex and extended problem domain. His or her actions may therefore not be in accord with the overall requirements of the problem. Another difficulty encountered may be that the sheer complexity of the problem defeats an individual's information-processing capability. When people are required to solve difficult knowledge-based problems under stress, a variety of behaviours are observed, e.g. 'vagabonding' may be exhibited, where he or she flits from one 'hot' issue to another without managing to solve any one problem. Another characteristic behaviour is called 'encystment', where a single problem is singled out for solution to the exclusion of all others. Error rates for these three types of behviour appear in Figure 4.2.

4.4 Human reliability assessment techniques

Human reliability assessment addresses three main areas: (1) the modelling and identification of potential human errors; (2) the quantification of the probabilities of these errors; and (3) the specification of error-reduction measures.

4.4.1 An approach to modelling human error in proceduralized situations

A framework to model human errors is given in Figure 4.3. This framework is intended to be used to predict errors in tasks such as maintenance or routine plant operations requiring skill- or rule-based processing. The first stage of the framework involves identifying the functions that humans are required to perform for the system to achieve its designed goals.

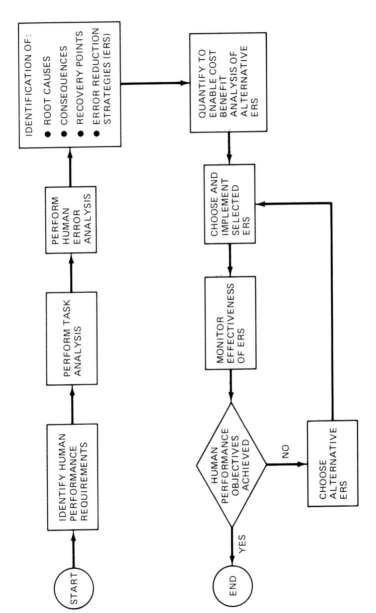

Figure 4.3 A systematic approach to analysing and optimizing human reliability

4.4.1.1 Task analysis

The tasks involved in performing these functions are then subjected to a systematic process of analysis called hierarchical task analysis (HTA) (Shepherd, 1984). In HTA, the goals that the operator has to achieve to perform the functions are successively broken down into the operations to be performed together with the plans which determine the ordering and timing of the operations.

The task analysis produces the following information:

1. Definition of the main task elements to be executed during the performance of the task. This is the primary path via which the operator is expected to achieve the task goals. Alternative paths are analysed in the error analysis.
2. Definition of the plans and the conditions which determine transfer of control between the levels in the goal hierarchy and within task elements at a particular level. This has particular relevance for the definition of training content, and the identification of certain types of errors (particularly slips).
3. Commentary on (1) and (2) with regard to their implications for procedures, training and equipment design. This information is subsequently supplemented by insights gained during the error analysis, and used in the error-reduction stage of the process. For an example of task analysis, see Table 4.2.

4.4.1.2 Human error analysis (HEA)

The next stage of the framework is concerned with predicting the errors. The objectives of the HEA can be summarized as follows:

1. Identify possible unrecovered error modes that could occur at each of the task steps documented by the HTA.
2. Identify the root causes underlying these error modes.
3. Postulate error recovery mechanisms.
4. Identify consequences of unrecovered error modes.
5. Develop recommendations for procedures, training and equipment design to minimize the probability of initial occurrence of errors and maximize their likelihood of recovery.

The process of identifying candidate errors that could occur at various stages of the task is achieved by the use of a computer program which identifies likely errors from the list of possible errors given in Table 4.3. The identification procedure depends on the answers to questions generated by the program regarding the characteristics of the task.

For each of the 'external errors' identified in Table 4.3, a number of root causes are possible. For example, for the external

Table 4.2 Typical entries in task analysis for human reliability assessment in a proceduralized situation

1 Task step	2 Inputs (cues for action)	3 Outputs (action)	4 Feedback	5 Error potential	6 System implications	7 Error recovery	8 Design implications
Addition of reactant B to Pan Q from charging vessel 4	Temperature in Pan Q greater than 200°C (chart recorder 6, trace 2)	Open valve 12B	Weight reading from charge vessel 4 indicates transfer complete. Temperature in Pan Q increases to over 250°C in 10 min	Could mistake valve 12A for 12B. (Close together on panel)	Hazard: 12A will cause reactant to be added to Pan Q causing dangerous exotherm. Temperature will rise to 450°C in 5 min	If 12A operated by mistake, weight reading from charge vessel 4 will not change. Rapid temperature rise in Pan Q. Operator can close valve 1B to prevent transfer if mistake detected within 2 min	Consider interlock. Clearly separate and renumber valves 12A and 12B. Insert procedural check for operator to monitor weight change in charge vessel and temperature rise in Pan Q

Table 4.3 External error types

Physical action
E1   Action too early
E2   Action too late
E3   Action omitted
E4   Action too much
E5   Action too little
E6   Action too long
E7   Action too short
E8   Action in wrong direction
E9   Right action on wrong object
E10 Wrong action on right object

Checks
E11 Check omitted
E12 Check on wrong object
E13 Wrong check on right object
E14 Check mistimed

Communication
E15 Information not obtained
E16 Wrong information obtained

Other
E17 Misalign

---

error mode 'action omitted' (e.g. the operator may omit to close a valve) the following root causes might be responsible.

1. *Place losing error:* The operator may be distracted and could lose his place in the procedure, thus omitting a step or steps.
2. *Assumption:* The operator could assume that someone else had closed the valve.
3. Short cut invoked: The operator may assume that he can get away with a short cut which does not require the valve to be closed.

A classification of twelve different types of root causes has been developed. As with the external error modes, a computer program is available which can be used to identify possible error root causes on the basis of questions put to the analyst.

4.4.1.3 Developing error reduction strategies
The error root causes identified in the preceding stage of the analysis are important inputs to the next stage, the development of error-reduction strategies. In the omitted valve closure example mentioned above, three possible error-reduction strategies are

suggested by a consideration of the error root causes. The place-losing error could be minimized by the provision of a checklist or by external checking by an independent operator. The likelihood of the 'assumption' error could be reduced by some local indication that the valve is not closed, e.g. by using a rising-stem valve. The likelihood of the 'short-cut' error could be reduced by demonstrating to the trainee that no short cuts are possible. Three broad error-reduction methods are usually considered: (1) the use of operating instructions or checklists; (2) the redesign of equipment; or (3) the use of training approaches.

Error recovery is also considered as part of the consideration of methods for error reduction. This is because many of the errors that could occur will be recovered (i.e. detected and appropriate action taken) before they have a significant impact on the system. The usual mechanism for recovery is that some subsequent phase of a task cannot proceed without an earlier step having been completed. The operator detects the earlier error when he or she attempts to carry out the later phase and (hopefully) corrects it before it affects the system. This form of recovery will, of course, be feasible only if there is direct feedback that an earlier failure has occurred.

The other issue considered at this stage concerns the consequences of an error. The consequences of failing to achieve particular goals will already have been considered during the task analysis phase, where it is used to determine which aspects of the task should be evaluated in fine detail. The consequences of errors at the level of individual task steps will also be used to decide if error-reduction measures are necessary or cost-effective.

4.4.1.4 Choosing and implementing error-reduction strategies

The choice of error-reduction strategies is ideally made on the basis of the quantification of error described in section 4.4.2. The reduction of error probability is evaluated as a function of the costs required for alternative strategies. Having chosen an appropriate approach to error reduction, the effectiveness of the approach then needs to be evaluated by data-collection exercises in which the incidence of errors and near misses is monitored. If the required human performance objectives are achieved, the analytical method has achieved its purpose; otherwise, alternative error-reduction methods need to be implemented.

4.4.2 Quantification of human error

The quantification of human error was the earliest historical aspect of human reliability assessment to be considered. In the development of the space and ballistic missile programmes of the early 1960s it was found that human errors were a major contribution to system failures. Analysts interested in systems-reliability assessment, using methods such as fault trees, analysed human error in this context in the same way as failures of pumps or valves. Thus, the main emphasis of early work in this area was simply to obtain numerical estimates for the probability of human errors for specified human functions, e.g.: What is the probability that the operator will operate the emergency fuel cutoff within 20 s of the onset of a fire? Since these functional requirements arose purely from the engineering analyses of the system, the need to postulate and model *alternative* operator actions that could affect the system, using a method such as that described in section 4.4.1, was not perceived as being important.

The most common method for human reliability quantification for whole tasks is called the decomposition approach. This involves breaking down a task into its component elements (usually task steps) and assigning error probabilities to these steps from a databank. The individual probabilities are then combined together to give the overall error probability for the task as a whole. The most widely used method of this type is the technique for human error rate prediction (THERP) (Swain and Guttman, 1983). One of the major problems in the area of human reliability quantification is the unavailability of error data on human performance in authentic situations. Even the databank used in THERP contains a high proportion of subjectively derived data. Techniques such as the success likelihood index method (SLIM) (Embrey *et al.*, 1983) use a form of structured expert judgement to extrapolate from a relatively small database of known error probabilities. They do this by developing a model which connects the probability of success (or failure) to the factors which are perceived to determine this probability, e.g. quality of training, time to perform the action, design of the equipment, etc. This approach provides a useful basis for cost-benefit analyses, since the effects of spending resources on error probability to modify these factors can be investigated.

A comprehensive review of human reliability quantification techniques is given in Embrey (1987). It can be said in summary that, although considerable resources have been devoted to the development of quantification techniques, real progress in this area is unlikely to be achieved unless serious attempts are made to collect data on human performance in real plants. This in turn is unlikely to occur unless a fundamental change in attitudes takes place, removing the connotation of blame from human error, so that individuals are prepared to provide this information.

4.5 Conclusions

This chapter has provided an overview of the field of human reliability assessment. It has shown that theory from areas such as cognitive psychology provides a theoretical basis from which errors can be predicted in certain types of task, particularly where skill and rule-based processing are concerned. Although major resources have been devoted to the development of quantitative methods for assessing error probabilities, these are largely limited by the absence of empirical data.

A number of areas are of particular significance in terms of future developments. The first is the development of systematic methods to model and predict errors in more complex situations involving knowledge-based processing, e.g. unpredicted plant emergencies. Some exciting advances have already been made in the use of artificial intelligence models to simulate the behaviour of nuclear power station operations in emergencies. These models could be used to predict operator performance for human reliability assessment purposes (Woods, 1987).

An increased interest in the effect of organizational and management variables on human reliability in the context of large systems, such as process plants and power stations, is another important trend. The modelling approaches described earlier are essentially 'bottom-up' in that they are concerned with the detailed interactions of the individual operator or team with the plant or equipment. Organizational and management factors obviously will have an impact on reliability at all levels, particularly via the effectiveness of the communication links, both horizontal and vertical, which exist in an organization (Bellamy, 1983).

Another area where human reliability considerations are starting to be evaluated is that of software reliability. Many of the cognitive psychology models which have been the source of techniques described earlier could be used to model the human contribution to software reliability.

A final area where human reliability considerations are likely to become increasingly important is that of product design. Developments in legislation for both consumer products and suppliers of industrial equipment have moved the onus of responsibility for safety to the designers of such products. The use of the systematic techniques described earlier could reduce the likelihood of accidents by modifying product designs on the basis of error analyses and also by assisting in the development of high-quality operating instructions.

The study of human reliability can be seen as a dynamic field which is expanding from its origins in large hazardous systems to a wide variety of new applications.

References

BELLAMY, L. J. (1983) 'Neglected individual, social and organizational factors in human reliability assessment'. *Proceedings, 4th National Reliability Conference, NEC, Birmingham*

EMBREY, D. E. *et al.* (1983) *SLIM-MAUD, An approach to assessing human error probabilities using expert judgement.* NUREG/CR-3518, Washington DC-20555

EMBREY, D. E. (1987) Human reliability: the state of the art'. *Proceedings, conference on human reliability in nuclear power*, London. IBC Technical Services, London

RASMUSSEN, J. (1980) 'What can be learned from human error reports?' In: K. D. Duncan, M. M. Gruneberg and D. Wallis (eds.) *Changes in Working Life*. John Wiley and Sons, Chichester

RASMUSSEN, J. (1982) 'Human errors: a taxonomy for describing human malfunction in industrial installations'. *J. Occupat. Acc.*, **4**, 311–335

RASMUSSEN, J. (1983) 'Skills, rules and knowledge: signals, signs and symptoms and other distinctions in human performance models'. *Inst. Electronic and Elec. Engrs Trans. Systems, Man and Cybernetics*, SMC-13. **3**, 257–266

REASON, J. (1987) 'The cognitive bases of predictable human error' in: E. D. Megaw (ed.) *Contemporary ergonomics*, pp. 21–36. Taylor and Francis, London

SHEPHERD, A. (1984) "Hierarchical task analysis and training decisions'. *Prog. Learning and Educ. Tech.*, **22**, 2, 162

SWAIN, A. D. and GUTTMAN, H. E. (1983) *Handbook of human reliability analysis applications.* Report NUREG/CR-1278, Sandia National Laboratories, Albuquerque

WOODS, D. D. (1987) *An artifical intelligence base cognitive model for human performance assessment.* Report NUREG/CR-4862, Westinghouse Corporation, Pittsburgh, Pa, USA

Chapter 5

Hazard awareness and risk perception

G. C. Simpson

5.1 Introduction

It would seem axiomatic that a high level of hazard awareness and a realistic assessment of the risks involved in a job are prerequisites of an effective safety programme, yet until recently these topics have received little more than passing reference in the literature on occupational safety. This may be partly because they are so 'obviously' an aspect of the safety equation that they have never merited specific discussion. After all, what is the purpose of safety training if not to make people aware of the hazards and risks of their work? Similarly, it may be partly due to a change in semantics and partly a question of nuances of interpretation. For example, the concept of accident-proneness, once so prominent in the safety literature, could equally be interpreted as a function of poor hazard awareness and risk perception rather than a fundamental personality trait. The classic paper by Hill and Trist (1953) which identified a relationship between accidents and absenteeism, and interpreted this on the basis of psychological withdrawal, could likewise be explained by failures in hazard and risk perception. In fact, as we shall see later, this interpretation has been vindicated in a subsequent study. Whatever the reason for this apparent paradox of central issues receiving little attention, it remains true that the systematic consideration of these factors in accident research is a relatively new development.

Obviously separate concepts, hazard awareness and risk perception are, in practice, so interlinked that they can be treated as effectively two sides of the same coin; the individual's personal awareness of danger is intrinsic to his workplace and working practices.

Recent studies in this area fall into two categories: (1) those associated with the theoretical concept of risk homeostasis; and (2) a more diffuse group which, for the want of a better phrase, can be called the pragmatic studies. This latter group have little in common other than their acceptance that more information on hazard and risk perception may be of value in the understanding of the causes of accidents.

5.2 Risk homeostasis

The theory of risk homeostasis (proposed by Wilde, 1982) states that people adjust their behaviour to changing circumstances in order to keep the objective risk essentially constant. Although this may appear at first to be a rather extreme idea, it is apparent in many everyday situations. Howarth (1987) provides an excellent example: we tend to drive faster on wide, clear roads and slower on the narrow, busy roads in town. Given that increasing speed increases the risk of an accident, it can be seen that effectively we increase the risk on the relatively safe roads while decreasing the risk on relatively dangerous ones. Although risk homeostasis has the merit of applicability to everyday experience and, as we shall see below, some experimental support, it has proved extremely controversial. The reason for this is that if the theory is true universally, any attempt to improve safety by engineering inevitably will fail; whatever changes are made, the theory predicts that people will simply change their behaviour in order to return to the level of risk that prevailed before the 'improvement' was introduced.

Two recent papers, Howarth (1987) and McKenna (1985), review both studies in relation to this concept and its underlying principles; it is therefore sufficient simply to provide a flavour of the debate. Most of the studies lending support to the theory have been related to road traffic accidents. The basic approach taken is essentially to predict the effect of a safety improvement prior to introduction and measure the effect afterwards. Any shortfall is then examined to see if there is any way in which it can be explained on the basis of behavioural adaptation. For example, Wilde (1982) cites the evidence that, while the introduction of traffic-lights changes the type of accident, it does not influence the level of casualties. Conybeare (1980) presented data which showed that the compulsory wearing of seat belts in Australia had no net effect. Although the number of occupant injuries went down, the number of non-occupant injuries went up. Rummer, et al. (1976) showed that on the introduction of studded tyres to improve grip in icy conditions, the benefit was negated – the drivers simply used the extra grip to drive faster. Each of these studies has been interpreted as supporting the concept of risk homeostasis. Several other papers in support of the theory have also been reported (see, for example, Wilde, 1984) and some examples have been suggested to show how the theory could be used to provide incentive schemes to promote safe driving (Wilde and Murdoch, 1982). However, there are also a number of papers which claim to refute the theory or, at the very least, lend no support in circumstances where supporting evidence might legitimately be expected. For example, Huddart and Dean (1981) have reviewed the evidence on the influence of highway modifications (which according to the theory should have no lasting effect) and shown that, while not all are successful in reducing accidents, many can be shown to have been extremely successful. Several studies of seat belt applications have been published which similarly do not suggest any form of behaviour compensation (e.g. von Buseck, *et al.*, 1979; Evans, Wasielewski and von Buseck, 1982).

Risk homeostasis has also been attacked in terms of the evidence available on the basic assumptions made. For example, people must be able, for the theory to make any sense at all, to make reliable assessments of subjective, and objective, risks inherent in a situation. There is, of course, much evidence that people are poor at risk estimation (especially the levels of probability associated with accidents) and of the variability of such estimations (see Slovic, Fischoff and Lichenstein (1981) for a review).

Both sides in the debate unfortunately leave themselves open to criticism from the 'opposition' in terms of methodological limitations and of selective use of supporting literature. For example, the supporters of risk homeostasis favour the emphasis given above on the results of the Rummer, *et al.* study. The detractors point out that, while drivers did increase their speed with studded tyres, this was marginal compared to speeds to be expected if the conditions had not been icy. It seems unlikely, therefore, that the theory will be vindicated or otherwise without considerably more investigation. The value of the theory itself may be limited, but there can be no doubt that the debate it has produced has highlighted the importance of risk perception in the understanding of accident causation, even if it has failed to show convincingly that it is the only issue of real significance.

5.3 Pragmatic studies

In an appendix to the classic report *2000 accidents*, prepared for the Robens Committee on Health and Safety at Work, Hale (1971) sets the scene for many studies of risk by examining what governs a person's appreciation of risk at work. His small, but well-conceived, study involved independent rating of risk factors in a light-engineering factory followed by detailed interviews with staff on the risks they saw in their workplace. Perhaps the most fundamental issue to arise was the importance of previous experience of an accident on the awareness of the associated risk. As Hale states:

> I asked the operator his, or her, opinion of the risk. The question I used was: 'How is it possible to hurt yourself doing this job?' From the replies it was clear that the question was usually treated as though it had been: 'How have you, or someone you know, been hurt doing this job?'

Hale goes on to cite several examples from the study which support this contention. He also derived a number of other influential factors. These included:

1. How frequently accidents associated with a given risk occurred; the more frequent the accident, the greater the risk awareness.
2. The seriousness of potential injury; the risks related to minor cuts and scrapes, considered 'part of the job', were rarely mentioned.
3. The personal characteristics of the individual; there was clear evidence that some workers had a much greater appreciation of risks than others doing the same work regardless of accident experience.

It is clearly apparent, therefore, that individuals do have different risk awareness. However, is there any evidence that such differences have any influence on accident likelihood?

Mention was made earlier of Hill's and Trist's extremely influential study in the early 1950s; a recent Belgian study not only replicated this work but also examined risk-taking as a possible alternative explanation. Verhaegan, *et al.* (1985) re-examined the relationship between accidents and absenteeism, with some methodological refinements, in two wire-mills in two different companies. The study confirmed Hill's and Trist's finding that there was a definite relationship between the two. They were also able to analyse on the basis of whether the accident victim had been 'active' or 'passive'. By 'active', they meant that the victim had actually been directly involved in the events which led to the accident, whereas the 'passive' victim had not been involved directly – 'an innocent bystander'. There was a preponderance of

Table 5.1 Absences of accident-free workers or workers with passive accidents compared with absences of workers with active accidents. (From Verhaegen, *et al.* (1985), Table 4)

	Number of absences over 6 months				Total
	0	1	2	3 or more	
No accidents or passive accidents	81	42	11	4	138
Active accidents	36	32	23	9	100
Chi-square = 19.00	degrees of freedom = 3				$p<0.001$

'actives' among those who also had high levels of absenteeism (Table 5.1). A further aspect of the study examined, using rating scales, the victims' attitudes toward their company and its safety rules. There was a significant difference between the 'active' and 'passive' groups, with the former being more negative both to the company and to the safety rules. Each of these findings could be explicable on the basis of Hill's and Trist's explanation of 'psychological withdrawal'; however, they were hoping for evidence to support a conclusion which would provide the basis for remedial action. This arose later in the study when risk perception was examined in more detail. Thirty of the subject pool were chosen, ten each from the three groups: 'active' victims, 'passive' victims and those who had had no accidents. The selection was random within each group. Each of the thirty went through a

Table 5.2 Mean scores on different rating scales obtained by ten workers with 'active' accidents, ten workers with 'passive' accidents and ten workers without accidents. (After Verhaegen, *et al.* (1985), Table 8)

	'Active' accidents	'Passive' accidents	No accidents	Significance
Risk taking	2.32^a	1.79^b	1.72^b	$p < 0.05$
Dangerousness of work	2.70^b	3.05^a	3.55^a	$p < 0.05$
Use of personal safety equipment	6.48	6.62	6.38	N.S.
Discomfort of personal safety equipment	1.74	1.94	1.80	N.S.
Positive attitude towards safety department	6.60^a	6.15^a	5.15^b	$p < 0.05$
Accident = chance event	3.80	4.24	3.85	N.S.

For a given rating scale the average scores with superscript *a* differ significantly from those with superscript *b* (multiple *t*-test).

series of interviews during which a number of questionnaires and rating scales were used. The topics covered included not only risk-taking but also the dangerousness of the work, the use of personal safety equipment, the discomfort of the safety equipment, attitudes to the safety function, and fatalism (are accidents simply a matter of chance?) The results showed (Table 5.2) that the 'active' victim group took significantly more risks than either of the other two groups (between which there was no significant difference). The 'active' group also returned significantly lower scores of their perceptions of the danger of their work than did the other two groups. Once again, there was no significant difference between the 'passive' victim and no accident groups. No significant differences at all were found on either of the questions relating to safety equipment, nor on the question of fatalism. However, an interesting result did arise on the attitude to the safety function where both the accident groups had 'better attitudes' than did the non-accident group. Clearly, direct involvement had been influential in forming positive attitudes!

Verhaegan, *et al.* drew the conclusion from this extremely complex and well-designed study that, while they had confirmed Hill's and Trist's findings, their additional areas of study had suggested a much simpler explanation – risk perception.

Ostberg (1980) investigated risk perception and work behaviour in Swedish forestry personnel. A booklet containing realistically drawn examples of felling situations was prepared to provide 45 paired comparisons. Over 700 subjects were used covering six groups: (1) fellers; (2) trainers; (3) safety officers; (4) forestry school trainees; (5) safety engineers; and (6) supervisors. The examples used in the booklet covered nine felling operations and the rating of relative risk across these operations was in very close agreement. However, there were interesting differences in the absolute risk levels as estimated by the groups studied. The trainers tended, as may be expected, to overestimate the risk. Of perhaps more concern, however, was the finding that the supervisors tended to underestimate the risk. This may, in part, explain the finding of Andriessen (1978) which suggested that the supervisors' attitude is particularly influential in safety motivation. Another of Andriessen's findings that safe working practices are often perceived as 'bothersome and annoying', finds support in studies of risk-taking behaviour in forestry operations. Pettersson, Sundström-Frisk and Werner (1980) interviewed forestry workers about deliberate risk-taking prior to a change from piecework to day rates; although it was apparent from the study that the workers knew the dangers and the importance of safe procedures, they considered them to be too time-consuming and therefore frequently ignored them. However, once the payment system changed so did the level of risk-taking.

Staying within the agricultural industry, Singleton, Hicks and Hirsch (1981) reported an interesting unpublished study of tractor-driving by Hirsch. Hirsch asked tractor-drivers to rate ten situations on how likely they thought a tractor would overturn in each situation. These ratings were then compared with the incidence of tractors overturning in these situations. For eight out of the ten situations, the subjective rankings agreed reasonably well with the objective. For the other two, however, there was a substantial discrepancy. For one – the tractor being hit by another vehicle on a public road – the drivers underestimated the risk substantially. They rated it tenth – the least likely of all the possibilities presented – whereas objectively it ranked fifth. The other discrepancy concerned the tractor rearing backwards while travelling up a slope. This they overestimated, placing it sixth, whereas objectively it ranked tenth – the least likely. Elsewhere in their paper Singleton, Hicks and Hirsch cite back injuries and chemical hazards as additional issues on which 'misperception or lack of awareness of hazards' are major factors in agricultural accidents.

A recent study in the mining industry (Rushworth, *et al.*, 1986) examined risk perception among men working on the maintenance and cleaning of coal-storage bunkers. Observations of work in a variety of bunkers, together with discussions with a range of staff, provided the basis on which to derive a representative range of risky behaviours. Descriptive scenarios were then written for each of the situations and an 'expert panel' used to place each situation on a risk probability scale, anchored to an everyday risk of known probability (being run over on the road). Workers from a range of jobs associated with bunker operations were then asked to rate the riskiness of each scenario using 'being run over' as a reference point. A total of twenty-two activities were rated by each subject: the first four were essentially practice questions, and the remaining eighteen the 'test' questions dealing specifically with bunker work. After the initial rating, each activity was discussed in detail regarding the risk of being run over as 'socially acceptable'. His ratings above this were termed 'risky', and those below, 'safe'. If he failed to mention any of the hazards, these were explained and he was given the chance to change his rating. For those he had rated as risky, he was also asked if he had seen anyone performing this activity and/or had done so himself and, if so, what he thought had been the reason for taking the risk. A series of multivariate analyses were carried out which produced a rather complex set of results, only a selection of which are described here.

The comparison of the various groups of workers provided some interesting results. The trainers appeared to have a good awareness of risk and this was confirmed by the relatively few changes they made in their initial ratings after discussion. The experienced colliers and craftsmen (who occasionally worked in bunkers) and the supervisors, appeared less well aware of the hazards and, despite a tendency if unsure to rate toward risky, frequently modified their rating after the discussion. The two groups of 'specialists' (used to working in shafts and bunkers) showed an interesting difference. The colliery teams followed the same pattern as that described above. The area teams, who covered a number of pits, were similiar to the trainers in that they rarely changed their ratings; unfortunately, however, their initial ratings often considerably underestimated the risk. The inexperienced group behaved differently again by rating activities as relatively more risky than the other groups, and being generally more likely to change their rating after discussion, as would be expected. If a change of rating after discussion is used as an indicator of poor hazard-awareness, a feel for the general extent of the problem can be obtained by looking at the overall number of changes across the eighteen activities. More than 10% of the sample changed their rating on ten of the eighteen activities and more than 30% on five of them. It was possible in this way to identify those activities which would merit more consideration during, for example, safety training.

Finally, it is worth mentioning the answers to the question: Why did the subjects believe that risky activities were used? Four reasons are clear:

1. Lack of awareness of hazards.
2. Expediency or convenience.
3. Job pressures – instructions, promotion prospects.
4. Overconfidence.

In a subsequent study by the same team (Graveling, *et al.*, 1987), the risk/hazard awareness assessment procedure was used in a bunker training course. The risky scenarios were presented to trainees at the start of the course and again at the end. The instructors in addition also completed the exercise at the end of the course (they were not given it at the beginning, to avoid biasing their training). Before-and-after comparisons were made to examine the extent to which the training had sensitized the trainees to the risks involved. The results generally indicated, using the instructors' ratings as the baseline, that the training had been reasonably successful. There were, however, indications which suggested that the procedure could be used both to identify individuals who required extra tuition on certain points, and as a feedback mechanism for the course itself by indicating specific topics which had implications that were not fully grasped by several of the trainees. Using risk perception in this way in safety training also seemed to suggest a useful yardstick on which to base the timing of refresher courses, especially for jobs such as bunker work, which is intermittent but involves complex safety issues.

The measurement of risk perception in safety training could be of considerable value as an indicator of comprehension of the course, if suitably refined to include objective baselines, such as the actual frequency of accidents as used by Hirsch in the study of tractor drivers.

5.4 Conclusion

Although the study of risk perception and hazard awareness as factors in accident causality has been far from systematic (and the only attempt so far to provide a theoretical framework less than convincing) there can be no doubt that the subject is of considerable importance. This becomes even more evident on examination of recent accident trends which increasingly emphasize factors such as: 'lack of due care and attention', 'failure to use safe working methods', 'human failings', etc. (see, for example, Health and Safety Executive, 1945; 1986). There is sufficient evidence in the studies quoted above to suggest that a greater understanding of risk perception will be valuable in the resolution of the human factors in the causes of accidents. Moreover, there is also the hint that studies of risk perception and hazard awareness help us not only to better understand accidents, but that they may also provide us with a new approach to amelioration.

References

ANDRIESSEN, J. H. T. H. (1978) 'Safe behaviour and safety motivation'. *J. Occupat. Accidents*, **1**, 363–376

VON BUSECK, C. R., EVANS, L., SCHMIDT, D. E. and WASIELEWSKI, P. (1979) *Seat belt usage and risk-taking in driving behaviour.* General Motors Research Laboratory (Report No. GMR-3116), Warren, Michigan

CONYBEARE, J. A. C. (1980) 'Evaluation of automobile safety regulation: the case of compulsory seat belt legislation in Australia'. *Policy Sciences*, **12**, 27–39

EVANS, L., WASIELEWSKI, P. and VON BUSECK, C. R. (1982) 'Compulsory seat belt usage and driver risk-taking behaviour'. *Human Factors*, **24**, 41–48

GRAVELING, R. A., MASON, S., RUSHWORTH, A. M., SIMPSON, G. C. and SIMS, M. T. (1987) *Utilization of accident data to improve safety in the human factors aspects of system design*. Institute of Occupational Medicine Report No. TM/87/16, Final Report on CEC Contract No. 7258/01/133/08. IOM, Edinburgh

HALE, A. (1971) 'Appreciation of risks at work'. In: P. I. Powell, M. Hale, J. Martin and M. Simon, (eds) *2000 accidents*. National Institute of Industrial Psychology (Report No. 21). NIIP, London

HEALTH AND SAFETY EXECUTIVE (1985) *Deadly maintenance – a study of fatal accidents at work*. HMSO, London

HEALTH AND SAFETY EXECUTIVE (1986) *Mines: health and safety*. HMSO, London

HILL, J. M. M. and TRIST, E. L. (1953) 'A consideration of industrial accidents as a means of withdrawal from the work situation. A study of their relation to other absences in an iron and steel works'. *Human Relations, 6*, 357–380

HIRSCH, A. (1981) 'A study of tractor drivers' perceptions of the risks in their work'. Unpublished MSc thesis (quoted in Singleton, Hicks and Hirsch, 1981)

HOWARTH, C. I. (1987) 'Perceived risk and behavioural feedback: strategies for reducing accidents and increasing efficiency'. *Work & Stress, 1*, 61–66

HUDDART, K. W. and DEAN, J. D. (1981) 'Engineering programmes for accident reduction'. In: H. C. Foot, A. J. Chapman and F. M. Wade, (eds) *Road safety: research and practice*. Pra, Eastbourne

McKENNA, F. P. (1985) 'Do safety measures really work? An examination of risk homeostasis theory'. *Ergonomics, 28*, 489–498

OSTBERG, D. (1980) 'Risk perception and worker behaviour in forestry: implications for accident prevention policy'. *Accident Analysis & Rev., 12*, 189–200

PETTERSSON, B., SUNDSTROM-FRISK, C. and WERNER, M. (1980) *Experience of the transition from piecework to fixed wage forms*. Forskningsstiftelsen Skogsarbeten, Ekonomi Report IE, FS, Stockholm

RUMMER, K., BERGGRUND, V., JERNBERG, P. and YTTERBOM, U. (1976) 'Driver reaction to a technical safety measure – studded tyres'. *Human Factors, 18*, 443–454

RUSHWORTH, A. M., BEST, C. F., COLEMAN, G. J., GRAVELING, R. A., MASON, S. and SIMPSON, G. C. (1986) *Study of ergonomic principles in accident prevention for bunkers*. Institute of Occupational Medicine Report No. TM/86/5, Final Report on CEC Contract No. 7247/12/049. IOM, Edinburgh

SINGLETON, W. T., HICKS, C. and HIRSCH, A. (1981) *Safety in agriculture and related industries*. Department of Applied Psychology Report No. AP106. University of Aston, Birmingham

SLOVIC, P., FISCHOFF, B. and LICHENSTEIN, S. (1981) 'Perceived risk: psychological factors and social implications'. *Proc. Roy. Soc. Lond., 376*, 17–34

VERHAEGEN, P., STRUBBE, J., VONCK, R. and VAN DEN ABEELE, J. (1985) 'Absenteeism, accidents and risk-taking'. *J. Occupat. Accidents, 7*, 177–186

WILDE, G. J. S. (1982) 'A theory of risk homeostasis: implications for safety and health'. *Risk Analysis, 2*, 209–225

WILDE, G. J. S. (1984) 'Evidence refuting the theory of risk homeostasis? A rejoinder to Frank P. McKenna'. *Ergonomics, 27*, 297–304

WILDE, G. J. S. and MURDOCH, P. A. (1982) 'Incentive schemes for accident-free and violation-free driving in the general population'. *Ergonomics, 25*, 879–890

Chapter 6

Workplace accidents: a case for safety by design?

G. A. Laycock

6.1 Introduction

Behaviour in a work system has been examined in Chapters 4 and 5. In this chapter the emphasis will be on factors at the workplace which may be considered to affect the workers' reliability in the system.

The notion of human error, to many, is that of 'accident-proneness', a concept proposed early in this century by the Industrial Fatigue Research Board (Greenwood and Woods, 1919); indeed, the cause of accidents is quite often attributed to 'human error'. The question needs to be asked: is human error actually the cause of an accident, or is it merely the symptom of a poorly designed work system? It is my understanding and experience that many of the accidents that are supposedly caused by human error actually occur because of a combination of poor equipment design, poor job design, or other influences on the person which have the effect of causing an error of judgement or behavioural change, resulting in an accident or unplanned incident. Such accidents thus are indicative of work systems or workplaces which suffer from an incomplete, or non-existent ergonomic input.

Published government statistics have listed 'human error' as the implied cause of accidents. For example, 11% of fatal accidents involving transport during the period of the study (1978–80) were caused by human error (Health and Safety Executive, 1982). If the category of causation attributed to vehicle and equipment design is included, the figure rises to 18%. I believe that the two cannot necessarily be considered separately since 'human error' is often produced either by poor equipment design, or because equipment is not designed in such a way that the effects of human error are minimized. These data give some indication of the assumed problem of 'human error'. I wish to attempt to redress the balance somewhat by focusing attention on some of the underlying factors which may contribute to errors of judgement or behavioural changes being made during the course of work. I want, in so doing, to draw attention briefly to the need to consider personal performance factors during accident investigations before moving on to examine philosophies of prevention.

6.2 Personal performance factors

There are many factors which affect how a person performs at work. These include:

1. *Selection procedures.* We may need people with particular levels of intelligence, previous education or special physical ability depending on the work involved. It is important to ensure that workers' abilities match the requirements.
2. *Training regimes.* In-depth and refresher training may be necessary to maintain the desired level of performance in normal and abnormal work situations.
3. *Stress.* This could result from family problems, e.g. an argument with a spouse, work pressure, personality clashes with workmates, etc. It is not exclusive to management.
4. *Health problems.* These could be either long-term, affecting the person's ability to carry out physical duties – the adverse effects of medication (e.g. drowsiness associated with certain drugs) or even the very short-term, – the 'morning after' a heavy drinking session, perhaps.

Far too little attention is given to any of these problems when accidents are investigated, which is unfortunate, since the individuals and management alike gain no benefit from being told that an incident was the result of human error. Far better to be able to qualify that an incident was caused by human error following a particular short-term problem. This allows some positive action to be taken.

This observation is still relevant even with the increasing sophistication of machine control. If we assume that a machinery accident can be considered to be caused by a mismatch between the actions of the operator and the machine design, it seems that two possible options are available to us when it comes to accident-prevention philosophies:

1. The traditional approach, which possibly receives too much emphasis '. . .if it's dangerous put a guard around it'.
2. The approach which often receives far too little attention, namely to ask: Why do it that way?

The first approach has been extremely valuable and remains so. My concern, however, is twofold.

1. There are many complex or sophisticated systems in use at the workplace where this philosophy will not be effective.
2. How often do we attempt to find out the root causes of an accident by asking why the operator took a certain course of action? If it is clear that it is dangerous to put one's hand into a machine, why do it?

A few case studies show how the first philosophy can be applied to accidents with a less than totally successful end-result and where better use of the second philosophy may have found the true cause of the accident.

6.2.1 Machine guarding

6.2.1.1 Removal of hazards by redesign

Incident   A man received lacerations to his right hand when it became trapped by a spiked feed roller on a chipboard-edging machine.

Circumstances   The spiked roller drew the edging strip from a coil and through a guide, in order to be applied to the edge of chipboard by pressure rollers. The operator was using the machine after removing a perspex guard to feed a new length of edging strip into the spiked roller drive. *The production pressure made him unwilling to stop the machine.*

Outcome   A previous accident at exactly the same point on the machine had resulted in the perspex guard being fitted. It was considered that even if it had been in position in this incident it would not have stopped access to the feed rollers and so was extended even further. The operator had been told that the machine must be stopped before carrying out this task.

Considerations   It was necessary to remove the guard to perform two routine operations: (1) occasional cleaning; and (2) feeding new strip into the rollers, which was done much more frequently than (1).
   A guide had been fitted to the machine in order to ease this latter task but was not very effective, with the result that the operators needed to guide the strip into the spiked rollers by hand. It would have been in the company's financial interest to have

investigated ways of keeping the machine running during this operation by redesigning the guide system so that it functioned properly. This would have allowed strip to be fed into the rollers with the machine in motion and without the operator having to be dangerously exposed. The guard could then have been extended to eliminate all access, and with a fail-to-safety interlock fitted would make for quick, safe access for cleaning without the requirement to remove it.

Approaching the problem in this way may actually have speeded up production and been acceptable to the operator, an aspect all too often overlooked in solutions to dangerous situations.

6.2.1.2 Crisp packet jaws

Incident   A machine minder operating a machine for sealing packets of potato crisps received severe injuries when his fingers became trapped in the sealing jaws.

Circumstances   Plastic was sticking to the hot sealing jaws and caused problems with poor sealing of subsequent packets. The machine stopped when he opened it to clear the plastic (*causing a reduction in his rate of production*). He found through experiencing this problem over and over again that he could reach underneath the guard up to the sealing jaws and pull the trapped packets free by lying on the ground (Figure 6.1). Having completed this operation this accident happened.

Outcome   A fairly simple solution was adopted, namely by fixing additional guarding around the bottom of the machine so that it was impossible to get underneath the guard. Although this was probably an essential modification as the machine did not comply with legislation such as Section 14 of the Factories Act 1961, it did not address the basic problem.

Considerations   The operator adopted this method of removing burnt plastic in order to avoid cutting his production capacity by stopping the machine, but, since he had been able to find a quick, easy way of tackling the problem himself, management did not know about the problem of the sticking jaws. It became clear shortly after fixing the additional guarding that a major rethink was necessary on the design of the sealing mechanism because production was suffering as a result of the excessive amount of machine downtime now that the only access was via the electrically interlocked cover.

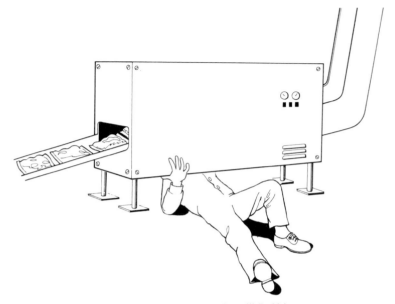

Figure 6.1 If there's a quicker way, your operative will find it!

If the question: Why did he do that?, had been more fully investigated at the time, the subsequent production problem may have been avoided. The vital role that proper machine fault reporting systems can play in preventing accidents and raising and maintaining production levels should not be overlooked.

6.2.1.3 Stopping machines with interlocked guards

Incident   A man was injured on a carpet tufting machine when his hand went into the nip between the bottom yarn feed rollers.

Circumstances   The injured man claimed he stumbled and, as he put out his hands, one became trapped. The company, however, suspected that he was incorrectly using a nearby interlocked guard panel as a means of stopping the machine remove to waste yarn instead of using the stop button.

Outcome   The action taken was to extend the existing interlocked guard panel.

Considerations   The company may have been correct in believing the operator was misusing the interlocked panel, but they did not follow-up their suspicions by analysing why it was necessary. The reason may have been, for example, that the stop button may have been too far away, or required a reset and restart procedure, whereas the interlocked panel may only have needed closing to allow the machine to continue operating. The use of the interlocked panel to stop the machine may therefore have been for convenience and to minimize downtime.

Further investigation may have been beneficial for the company. The action they took may prevent a similar injury, but did it really remove the cause of the accident?

6.3.2.4 A different approach to an 'unguardable hazard'

Incident   A man suffered a fractured finger while operating a machine that applied plastic coating and an identification code to copper pipe.

Circumstances   Individual pieces of pipe were formed into a continuous length utilizing plastic sleeves. The pipe then passed through a machine and the individual pieces were then separated again once the process was complete. This separation should have been done at a position clear of all ancillary equipment but this operator began separating the pipe near a pair of non-powered rollers through which the pipe had to pass. The rollers were friction-driven by the pipe and his finger became trapped in the in-running nip formed between the pipe and rollers.

Outcome   The operator's reason for cutting the pipe near the rollers was not determined and the conclusion drawn was that an adequate guard could not be designed to prevent other similar accidents and as a result, no action was taken.

Considerations   A different approach could, and should, have been taken. No one questioned why the operator chose to cut the pipe near the rollers. It may have been easier, more convenient, or quicker than the standard procedure, or the reason may lie with the previous operator who may have passed on the unsafe practice. If it was any of the first three reasons it may be that the procedure for cutting the pipe needed rethinking to allow the machine to be operated more efficiently. This may prevent further accidents, and may possibly improve quality and overall performance.

We see in this case the common approach to accident investigation and prevention: what caused the injury, and how do we fix a guard around it to stop access? By asking what the operator was doing and why he was doing it we may discover the real reason why an accident happened and not just where it occurred.

The worker's schedule was a further aspect that should have been examined. The accident took place at 1.40 a.m. on a night shift which had begun at 10.00 p.m. the previous evening. If this was the first or second night shift worked, his or her circadian cycles may have been confused and therefore he or she was not performing to the best level. Such considerations are rarely acknowledged by accident investigators and, although there is still considerable debate on the effects of rotating shifts on the ability to perform efficiently, perhaps more account should be taken of these factors.

6.2.2 Plant and equipment design

6.2.2.1 Cyclohexylamine release

Incident A worker incorrectly positioned a switch operating a diverter valve while operating a chemical batch processing plant and chemical fumes were released into the factory.

Circumstances Two reactors, 1 and 2, were used alternately to produce a batch product to be further processed within a centrifuge holding vessel 1. The feed of cyclohexylamine to reactors 1 and 2 was controlled by using a pneumatically activated diverter valve (Figure 6.2).

Reactor 1 would be charged with cyclohexylamine and various other chemicals whilst the charged reactor 2 was being processed. When the reaction within reactor 2 was complete, the contents would be transferred to holding vessel 1 and reactor 2 would then be recharged with cyclohexylamine and other chemicals. During this operation the contents of reactor 1 were processed and would then be transferred to holding vessel 1, reactor 1 then being recharged, and so on.

On the day of the incident, reactor 2 was being charged and reactor 1 contained a complete batch ready for transfer to holding vessel 1. The process operator charged reactor 2 with water and then proceeded to charge with cyclohexylamine. He made a mistake with the position of the switch operating the pneumatic valve and wrongly charged the already full reactor 1.

A procedure that had been used on other similar occasions was carried out; unfortunately, a second mistake was made during the recovery procedure and cyclohexylamine fumes were released into the plant.

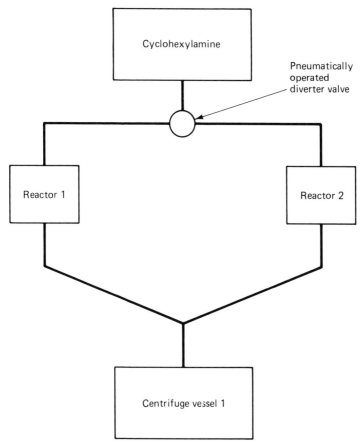

Figure 6.2 Cyclohexylamine fume release at a chemical batching plant

Outcome   The prime cause of this incident was considered to be operator error and it was considered vital that the batch sheet – a document listing the steps during normal operation – should be improved. The one-page document was replaced by an extended edition of four pages and, additionally, a three-page operator test questionnaire was devised for regular use. Nothing was done about the original cause of the incident – the wrong setting of the valve due to its design – even though this mistake had been made before.

Considerations   The improvement of the batch sheet may have been needed, but it did not help to prevent incorrect vessel charging occurring again through a mistake in setting the control for the valve supplying cyclohexylamine to the two reactors. Simple redesign of the valve control could remove the possibility of this type of error being made, hence making the recovery procedure unnecessary and helping to simplify all the written instructions rather than making them more complex.

6.2.2.2 Acid burns from reading sight glass

Incident   A probe on a measure pot was malfunctioning and a fitter was called to repair it. He started work on the wrong one – a measure pot full of concentrated sulphuric acid. On loosening the probe he was sprayed with acid on his face and arms and received severe burns.

Outcome   It was felt that the permit-to-work system in use needed revision and the fitter was also reprimanded for not wearing protective clothing; this latter aspect was considered of major importance.

Circumstances   There were two identical measure pots, one of which had performed without any problems, but the other needed frequent attention to the probe (twenty times in 3 years). On this occasion it was the non-troublesome one which was malfunctioning. The permit-to-work sheet made out by the charge-hand failed to state this and the fitter assumed the faulty valve to be the one that had caused problems before and as a consequence began working on the measure pot still in use and full of acid. On loosening the probe, acid began to leak out and the fitter decided to check the pot was empty by looking at the sight glass. As he could not see a liquid meniscus he assumed it to be empty. Unfortunately it was so full that the liquid meniscus was beyond the top of the transparent section of the sight glass and therefore out of view (Figure 6.3). He proceeded to remove the probe and was sprayed with acid.

Considerations   Slack liaison between the charge-hand and the fitter and the lack of protective clothing contributed to the severity of this accident, but a further important factor was the design of the sight-level glass. It would be capable, if redesigned, of indicating a completely full pot; the fitter would be able to check this instrument in future and would thereby be warned of danger.

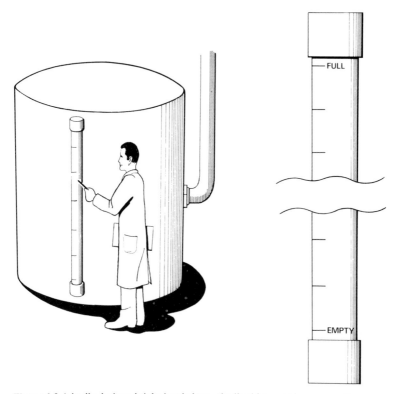

Figure 6.3 A badly designed sight-level glass – the liquid cannot be seen easily

6.2.2.3 Dumper truck pedal pressures

Incident   The driver of a 3 t dumper truck received fatal injuries on a construction site when vehicle overturned into a 3 m deep trench. His foot had slipped off the clutch pedal and he lost control.

Circumstances   One notable factor was the very heavy pressure which was required to depress the clutch pedal. When subsequently measured it was found to require a force in excess of 45 kg.

A pedal which requires a person to almost stand on it, as was the case in this accident, is clearly inappropriate. The necessity of using such force, especially while wearing muddy boots in adverse weather and over rough ground makes it more likely that the foot can slip off the pedal and reduces the amount of fine control.

Considerations   There is no British Standard which applies to dumper truck design, but BS 5528:1981 *Operator controls on excavators used for earth moving* which is the only closely related standard recommends a maximum operating force of less than 45 kgf for foot pedals. Even so, the situations where this maximum force is applicable are not specified. In any case, sitting in a large, reasonably stable, slow-moving excavator is clearly different to a small, more mobile dumper truck bouncing over rough ground.

Various publications in the field of human factors and ergonomics textbooks recommend maximum forces of the range 4.5 to 27 kgf. The *Handbook* edited by Galer (1987) does not quote forces but says:

> A control should require the application of a large force only if it falls into one of the following categories:
> - those used in an emergency as, for example, in the event of a power failure;
> - those which are used only occasionally and where power operation is considered unnecessary;
> - those which are operated by hand-tools during maintenance work.

It seems appropriate to suggest that the force levels of the order quoted in this text should be adopted as operational standards for such equipment.

6.2.2.4 Workplace adaptability

Incident   A female press operator slipped from a timber baulk on which she was standing and sprained her ankle.

Circumstances   The incident occurred in a metalworking shop where various bench-mounted tools were grouped together for a fairly simple process to be carried out on small workpieces. The operator given the job was a woman of average female stature who quickly found the workbench and the small press fitted to it were far too high for her to use comfortably. A large piece of redundant timber similar in size and shape to a railway sleeper was handy and quickly adopted as a convenient working platform (Figure 6.4).

Considerations   Having to balance on such a small area (in high-heel shoes) made it inevitable that she would eventually slip from her precarious perch. While the type of footwear worn was inappropriate, the proper solution should have been to provide a purpose-made raised platform to allow her to carry out work properly.

Figure 6.4 Workplace improvization – timber baulk used as a platform

6.3 Summary

Accident investigation in industry is carried out for several important reasons. These include the compliance with legislation (which does not concern us in this chapter) and the identification of the cause in order to take the appropriate action wherever practicable so that recurrences can be prevented.

The traditional approach – of identifying mechanical or electrical hazards and applying appropriate guarding techniques to prevent access – has been adequte for many years but with the increasingly sophisticated technology found in industry today this approach is having to be modified. It is not just at the complex level that a different approach is needed but also when examining the simpler, more commonplace accidents that occur hundreds of times every day.

If more emphasis could be placed on finding the reasons why the injured people took the action they did, we would be able to achieve a greater degree of protection.

Safety by design does not just apply to existing equipment but also to factory areas where new plant layouts are being installed and to transport systems on industrial and commercial sites, as well as to new equipment.

New equipment presents additional problems for the safety officer, engineer or other persons responsible for ensuring a safe working environment. Section 6 of the Health and Safety at Work, etc. Act 1974 has imposed duties on designers, manufacturers, importers, suppliers, erectors and installers since 1 April 1975 in order to ensure, so far as is reasonably practicable, that articles and substances for use at work are safe and without risks to health when properly used. Twelve years on, the principles put forward in Guidance Note GS8 (1977) as a means of complying with Section 6 are ignored far too often.

The answer to many of these problems must be the more widespread ergonomic education of those who influence the design of individual elements or complete systems.

If there is one important principle that designers, engineers or safety professionals should follow, it is that in any work situation we have tremendous scope to design the working environment to suit the human being but very limited scope for redesigning the human being to fit the work environment.

References

BRITISH STANDARDS INSTITUTION (1988) *Operator controls on excavators used for earth moving.* Milton Keynes.

GALER, I. (ed.) (1987) *Applied ergonomics handbook,* 2nd edn. Butterworth, London

GREENWOOD, M. and WOODS, H. M. (1919) *The incidence of industrial accidents upon individuals with special reference to multiple accidents.* Industrial Fatigue Research Board, Report No. 4, HMSO, London

HEALTH AND SAFETY EXECUTIVE (1982) *Transport kills. A study of fatal accidents in industry, 1978–1980.* HMSO, London

HER MAJESTY'S STATIONERY OFFICE (1974) Health and Safety at Work, etc. Act. HMSO, London

HER MAJESTY'S STATIONERY OFFICE (1961) The Factories Act. HMSO, London

HEALTH AND SAFETY EXECUTIVE (1977) *Articles and substances for use at work.* Guidance note GS8. HMSO, London

Part 3

User-centred design

Chapter 7

User-centred design – an ergonomist's viewpoint

S. T. Pheasant

A little neglect may breed mischief, . . . for want of a nail, the shoe was lost; for want of a shoe the horse was lost; and for want of a horse the rider was lost.

Benjamin Franklin, *Poor Richard's almanack*

It has long been an axiom of mine that the little things are infinitely the most important.

Arthur Conan Doyle, 'A case of identity', *The adventures of Sherlock Holmes*

The worst peace-time maritime disaster, involving a British ship, since the loss of the *Titanic* in 1912, took place on 6 March 1987. The cross-channel ferry *Herald of Free Enterprise* capsized just outside Zeebrugge harbour. Of the 459 (or more) passengers and 80 crew on board, at least 189 died; the exact figure remains a little uncertain. It was not possible to attribute this disaster to any single cause, since a complex set of contingent circumstances were involved. In comparison with other types of vessel, a roll-on roll-off car ferry is unstable by reason of her top-heaviness. The large undivided empty spaces, close to the water line, make her riskier still. She is a very unforgiving type of vessel. On this particular day, the combination of a high tide at Zeebrugge and the design of the port's loading ramp resulted in the *Herald* leaving port with her bows low in the water; regrettably, her bow loading doors had been left open, and the resulting inrush of water caused her to keel over in a matter of seconds. The captain and chief officer, who were held responsible for this 'error of omission', lost their licences – the captain for 1 year and the chief officer for 2 years. The standard procedure for closing the doors was nothing like foolproof; in fact, given the pressure on the crew to get the fastest possible turnaround time in port, it might be regarded in retrospect as a formula for disaster. But from the ergonomist's viewpoint, the most remarkable feature of the whole sorry affair was that there was no warning light on the bridge (nor any other kind of display or indicator) to tell the captain whether the doors were open or shut.

The absence of a warning light stands in stark contrast to the sophistication of the computer system in the engine room, which monitors some 256 different parameters in order to optimize the ship's operational performance, such as her speed and fuel economy. Loading-door warning lights are now mandatory.

A ferry captain working for the same line had requested that such lights be installed some years earlier. Had management heeded, instead of derided, this perfectly reasonable request the tragedy would probably have been averted. The cost of installing such a light? Probably in the order of £100 or, in nautical terms a ha'p'orth of tar. The cost of a car ferry and 189 lives? Decide for yourself according to whatever criteria you think appropriate.

7.1 Defects in the man–machine interface (MMI)

7.1.1 Acute failures

The circumstances surrounding the loss of the *Herald of Free Enterprise* are by no means unique in the annals of man-made disasters. Many catastrophic failures show points of similarity to the extent that a relatively consistent pattern of factors begins to emerge. Complex systems almost inevitably fail in a complex way; although the failure may be attributable to a single *proximate cause* (e.g. the bow doors being left open), this will generally be the consequence of a number of *antecedent causes,* and there will be other *contingent events* or *contributory factors.* Within this web of circumstances we will commonly encounter a 'human error' (i.e. a situation in which a person engaged in the operation of the system, was found to have left undone those things which he [or she] ought to have done, or done those things which he ought not to have done' (to quote the General Confession in the *Book of Common Prayer*). But if we back track from this point along the chain of antecedents, often we will find a defect in the design of the man–machine interface (MMI) i.e. the imaginary surface across which information is transmitted between the technical system and its human operator by means of displays and controls. The defect on the *Herald of Free Enterprise* (e.g. the absence of a warning light) renders the operator's task very much more difficult and, hence, predisposes him to errors of omission or commission.

Complex, technically sophisticated systems are characteristically vulnerable to the effects of human error. However, as discussed in

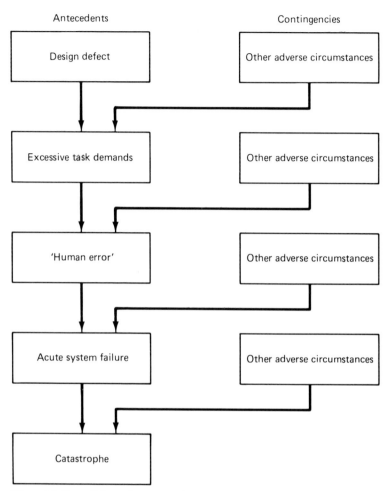

Antecedents

Contingencies

Design defect

Other adverse circumstances

Excessive task demands

Other adverse circumstances

'Human error'

Other adverse circumstances

Acute system failure

Other adverse circumstances

Catastrophe

Figure 7.1 The aetiology of a catastrophe

previous chapters, the diagnosis of 'human error' is uninformative and fundamentally unhelpful. It tells us virtually nothing about how we might avoid such problems in the future – unless we can think of ways of selecting and training operators to have perfect memories and full concentration. (An improvement in the human animal which properly belongs to the realms of science fiction.) Neither is it much help (people being what they are) to prosecute the erring operator and take away his licence (*Herald of Free Enterprise*) or send him to a labour camp (Chernobyl). But if we are able to identify those deficiencies in the MMI which overload the operator, we have the beginnings of a practical programme for dealing with the problem. And the surprising thing is that these design defects often concern very simple things, such as displays which are not present, controls which are difficult to operate and so on, so that with the wisdom of 20/20 hindsight, we are inclined to say: 'That's just common sense.' The defects can often be corrected by technical means which are both cheap and relatively unsophisticated – the ha'p'orth of tar or the horseshoe nail, in fact.

We might call this pattern of circumstances the Zeebrugge/Harrisburg syndrome (Pheasant, 1988). Its principal features are summarized in Figure 7.1. I do not wish to claim that this syndrome is implicated in all catastrophic failures, but it is possible to demonstrate (on a case-study basis at least) that it occurs sufficiently frequently to be both a major cause for concern and the most practicable approach to dealing with the problem. Let us therefore consider some more system failures, looked at from the ergonomist's standpoint. The examples I have chosen range from the catastrophic to the mundane, but the problems they illustrate are fundamentally similar.

7.1.1.1 Air crash at Orly airport, Paris

One of the worst air crashes of aviation history occurred on 3 March 1974, when a DC 10 airliner crashed shortly after taking off from Orly airport just outside Paris, killing 346 people. It was at first believed that the disaster had been caused by a terrorist bomb, but it was shown subsequently that the plane's cargo door had blown open, causing a rapid decompression in the cabin and the floor to collapse on to vital control cables and hydraulic lines. It was proved that the fastening mechanism of the cargo door had serious design defects and was extremely difficult to operate. It was concluded that the baggage handler at Orly (who was working under pressure due to industrial action by air-traffic controllers in London) had failed to close the door properly.

7.1.1.2 Petrol caps for Metros

Late in 1986, Austin–Rover recalled a number of Metros and supplied their owners with new petrol caps. The original design of cap had been difficult to fit correctly. If the tank was full, and the cap was fitted improperly, it was possible for fuel to leak over the rear wheels, causing the vehicle to go out of control.

7.1.1.3 Air crash at O'Hare airport, Chicago

In 1968, a Convair 580 aircraft crashed at Chicago's O'Hare airport with the loss of 27 lives. Subsequent investigation suggested that a pushbutton control could have been activated accidentally by the pilot's elbow, causing incorrect information to be displayed on the instrument panel, leading the pilot to make incorrect decisions, leading to the crash. A court was sufficiently convinced by this theory to award the victims' families $3.5 million in compensation.

7.1.1.4 Bad placing of headlamp switch
The headlamp switch on a certain British car design is mounted on the left-hand side of the steering column. It measures 21 mm square, and operates 'up' for 'on', contrary to the British and European conventions for switches. It is easy for the driver to operate the switch with his knee as he leaves the vehicle. The switch also lights the instrument panel, but this is not visible in daylight; hence, the driver is unlikely to notice that his headlamps are on. A flat battery may easily result. (I speak from experience.)

7.1.1.5 Collisions at automatic level crossings
After nine people had been killed on a gateless crossing at Lockington, Humberside in 1986, a report of the study commissioned to investigate safety at automatic level crossings concluded that collisions between trains and motor vehicles were almost always due to the 'failure of drivers to observe the signals'. Among the recommendations of the report was that the voltage of the crossing warning lights should be increased to make them more visible, and also that they should be shielded to reduce the effects of sunlight.

7.1.1.6 Doors closing in buses without warning
In Walthamstow, East London, a 79-year-old woman was dragged along the road with her leg trapped in the folding door of a bus and subsequently died in hospital. An auditory alarm, which should have warned the driver that the doors were not properly closed, had been disconnected.

7.1.1.7 Aircraft wandering into forbidden airspace
On 31 August 1983, a Korean Airways Boeing 747 (flight number KAL 007) wandered 300 miles off course into Soviet airspace on a flight from Anchorage to Seoul and was shot down by fighters. All 269 passengers and crew were killed. The most likely explanation seems to be that: (1) either the longitude of Anchorage was entered into the keypad of the automatic navigation system as 139 W rather than 149 W; or (2) that the autopilot mode selector was left in 'heading mode' rather than 'navigation mode'. These

two conditions are at adjacent settings, about 35° apart, on a five-way rotary selector, which is not particularly conspicuous on the cockpit's extremely complex console.

The list of catastrophes in which failures of the MMI were significant causative factors could be extended almost indefinitely. At the Three Mile Island nuclear plant, operators took an incorrect course of action because a visual display was giving them incorrect information. Design defects contributing to the great New York City blackout of 1977 included a display which gave ambiguous information and an alarm which was not in the same room as the operator who should have acted upon it; the final system failure was precipitated by the operator turning a master switch in the wrong direction. At Bhopal there were certain critical warning displays which were not working. The *Torrey Canyon* ran aground because her captain failed to switch from manual to automatic control while attempting a very difficult passage through the Scilly Isles in order to meet a deadline.

To summarize: we are dealing here with situations in which the safety or operational effectiveness of complex systems have been compromised by defects in the design of the MMI and which have rendered the systems difficult or impossible to operate, i.e. high technology systems which have failed because of low technology design problems. The ergonomist would argue that the elimination of these design defects is the best way of making such systems safer. A brief checklist of some of the more common deficiencies of interface design is given in Table 7.1.

Table 7.1 Some common deficiencies of interface design

(A) *Displays*
    A1 Display not there
    A2 Display gives incorrect information
    A3 Display unreliable (therefore ignored by operators)
    A4 Display not visible/conspicuous
    A5 Display not legible (graduations, labelling, lighting, etc.)
    A6 Display not intelligible (ambiguous, unclear, etc.)
    A7 Display contrary to stereotype or convention
    A8 Confusion between inadequately differentiated displays
    A9 Display of irrelevant information leads to confusion

(B) *Controls*
    B1 Control not accessible
    B2 Control too difficult to operate
    B3 Control operable accidentally
    B4 Control contrary to stereotype or convention
    B5 Control not compatible with displays
    B6 Confusion between inadequately differentiated controls

7.1.2 Chronic failures

The cases we have considered so far are all examples of *acute system failures,* i.e. they have occurred relatively unexpectedly, over a time-scale which can be measured in minutes or hours. Deficiencies of interface design may also result in *chronic system failures,* in which the systems concerned never actually suffer a catastrophic or unexpected breakdown but fail to live up to the expectations of their builders, on an ongoing basis, over a time-scale measurable in months or years. (I have borrowed this terminology from medicine, which is the oldest profession dealing with the failure of complex systems.)

Chronic dysfunction of this kind seems to be a characteristic of new technology products, commonly by reason of their incomprehensibility for their users; to use a rather appropriate bit of jargon, the interfaces are not sufficiently *transparent.* The users do not fully comprehend how the system works, or what it can do, so they fail to exploit it to its full advantage. The programmable timing devices, now commonly provided on electric ovens, are cases in point: experience suggests that they are scarcely ever used, even in situations where they would make life very much simpler for the user.

Williams and Crawshaw (1986) described a series of studies of domestic central heating systems. Their overall conclusion was that users characteristically failed to get the degree of warmth and comfort they wanted and as a consequence paid higher fuel bills than necessary. But this was not through any technical or engineering deficiencies in the systems, so much as a lack of comprehension of the systems by their users. In one study, the owners of programmable central heating systems were posed certain simple problems such as how to set up the system to keep them warm while they watched the late-night movie, or how to get lots of hot water quickly upon returning from a holiday. Less than half of the subjects were able to solve these problems correctly, using a programming device of the type fitted in their own homes (Bartram, Crawshaw and Williams, 1985). Dale and Crawshaw (1983) found considerable room for improvement in the timer and thermostat controls of these systems. The older electro-mechanical controllers commonly involved a disc of 65 mm or less in diameter, marked with a 24 h clock. Around its edge might be four tiny tappets which had to be set in place to select the times when the heating was switched on or off. Labelling was often small and difficult to read, especially in the poorly lit locations where these devices are commonly installed, and at times its ambiguity would stretch one's credulity. The more modern push-button microchip controllers were not always an improvement. A single digital display was used in some cases in different modes to show several different variables, the significance of which would almost certainly escape the less able user.

The most notorious problems of incomprehensibility are encountered with office technology in general, and with word processors and personal computers in particular (I must personally admit to finding photocopiers increasingly baffling). It is, perhaps, something of a counsel of perfection to demand that the workings of a sophisticated software package should be entirely self-evident in the hands of a novice user; and since personal computers generally come with instruction manuals, the idiosyncrasies of the software designers might perhaps be viewed with a certain amount of tolerance. (There is a school of thought which holds that it does not matter too much if the dialogue is a bit peculiar, provided that the manual explains it adequately.) But experience suggests that manuals are sometimes even less comprehensible than the software they purport to explain.

I have not yet come across any detailed empirical evaluations of computer manuals as such. (I find this remarkable – perhaps accounts exist outside the public domain.) But the following investigation, reported by Conrad (1962) should serve to illustrate the point. Conrad's laboratory had just been supplied with a new telephone extension system. He found its instruction booklet difficult to understand, so he devised a simple experiment to investigate the matter. Four matched groups of subjects attempted to transfer an outside call to another internal extension, without the help of the switchboard operator (this is by all accounts easy when you know how). The first group, who were supplied with the original instructions, had a success rate of only 20%. The second group received exactly the same instructions, except that they were retyped for better legibility, with the result that about 35% of this group were successful. The other two groups received instructions which had been rewritten in various ways for better intelligibility. They both achieved success rates of 70% or more. It is of some interest that a substantial improvement of intelligibility could be achieved by typographic means alone. Text which is difficult to read is necessarily difficult to understand. This is presumably because the effort involved in distinguishing the letters and words, interferes in some way with the higher-level task of determining their meaning. There is also a surprisingly large number of people who have uncorrected visual defects. However, the most striking feature of these experimental results, is the magnitude of the difference (350%) between the best and worst conditions. The moral of this story? We may regard the instruction manual as an extension of the MMI – as another display, in fact. We find once more that the performance of a technically sophisticated system, the research and development for which was presumably a very costly process, has been compromised by a lack of attention to the details of interface design, details which could be remedied easily and cheaply.

I recently came across a report of the misfortunes of a young man who tried to deposit his weekly wages for safe-keeping in the cardcash machine in his building society. It was just getting dark and he had not used the machine before. He found himself presented with two similar horizontal slots, and placed his money in the larger one. This turned out to be a gap behind the hinge in the machine's door. His money was stolen during the course of the night. The boy and his father subsequently complained (quite rightly) that the design of such a machine should be made 'foolproof'. Taken literally, this would appear to mean that the machine should be so designed that even a fool could operate it. If my recent experiences with photocopiers are anything to go by, I would include myself in this category of operator.

7.2 Design ergonomics

I have so far presented a gloomy picture of things: a world of man-made disasters, systems failing chronically and inscrutable machines which consume your cash. Many people really do see modern technology in these terms, a condition called *technophobia*. It is often easier to be critical than constructive, and in writing about ergonomics this is particularly the case.

Ergonomics tends to be one of those things which is most conspicuous by its absence. When clothes fit us comfortably and well we do not notice them; we only become aware of them when they are too tight or they hamper our movements. The answer to these problems, and the only antidote to technophobia, lies in a radical reappraisal of the way we build objects for human use.

Ergonomics began as an applied science concerned with people at work. Since workers generally use tools or machines, ergonomics came to focus particularly on the MMI. Tools and machines must be designed, so ergonomics evolved a philosophy and methodology of design. The ergonomic approach to design may be conveniently described by the term 'user-centred'.

7.2.1 User-centred design

The *principle of user-centred design* may be stated thus: 'If a product, an environment or a system is intended for human use, then its design should be based upon the characteristics of its human users' (Pheasant, 1986; 1987).

User-centred design has a number of features. These are not to be regarded as either necessary or sufficient conditions, defining the ergonomic way of doing things, so much as characteristics which are often present. The word 'product' will be used as a convenient shorthand, to denote any artefact intended for human use.

1. *User-centred design is empirical.* It is based on direct observations of human beings and their behaviour, supported by systematic investigations of human experience. It is generally distrustful, both of grand theories and intuitive judgements.
2. *User-centred design is iterative.* It is a cyclic process in which a design phase alternates with a phase of empirical analysis and evaluation.
3. *User-centred design is non-Procrustean.* It aims to modify the product to fit the user, rather than vice versa, and is concerned with people as they are rather than as they might be. (Procrustes is a figure in Greek mythology, who chopped off portions of his guests so that they would fit his bed more exactly.)
4. *User-centred design takes due account of human diversity.* It attempts to achieve the best possible match for the greatest number of people so far as is reasonably practicable within constraints of cost, etc. (A satisfactory match may be said to have been achieved when an individual is able to use the product concerned efficiently, safely, comfortably, etc. as defined by appropriate operational criteria.)

What do we mean when we say that a product is *'ergonomically designed'*? Does the term have any legitimate usage, or is it just advertisers' hype? Sometimes regrettably, the latter would seem to be the case: a year or two ago, ergonomists were surprised (and irritated) to see a certain make of car described as 'ergonomically designed, yet comfortable', and very recently to see a certain range of office seating described as 'designed with ergonomics in mind, yet simple to operate'. In both cases, the copywriter seems to have got the wrong end of the stick.

I recently came across a description of 'ergonomic pasta', which was designed for efficient straining and sauce retention – a case of fitting the noodles to the user?

Both the car and pasta examples suggest that 'ergonomically designed' means the same thing as 'easy to use', 'functionally efficient' or 'fit for purpose'. This usage is slightly off-centre. It makes ergonomics both broader than it really is (by encompassing all aspects of function) and narrower (by ignoring other ergonomic criteria). Functional efficiency and fitness for purpose are necessary, but not sufficient, conditions for ergonomic design.

7.2.2 User trials

In order to substantiate the claim that a particular product was ergonomically designed, we might reasonably look for evidence that it was, indeed, appropriately matched to the mental and physical characteristics of its users. We might, for example, conduct a user trial, i.e. an experiment in which suitably chosen subjects were asked to perform a representative range of tasks using the product concerned. Bartram's study of central heating controls and Conrad's experiment wth the telephone switchboard are both good examples of user trials. There is at least one case in which the detailed specifications for a user trial form the subject of a British Standard, BS 6652 *Packagings resistant to opening by children.* However, here we are concerned with demonstrating that a certain population of users is unable to perform a particular task.

The most obviously relevant outcome measure to take in any user trial is success or failure in the task. Thus, it would be interesting, for example, to see supposedly 'user-friendly' computer systems subjected to this sort of test. The subjective impressions and experiences of the users are also a matter for concern and there are some ergonomic criteria, most notably comfort, which can only be evaluated in this way. It is sometimes possible to back-up the users' reports of their subjective experiences, with objective measures of the physiological loading to which they are exposed. Life and Pheasant (1984) describe an experiment, concerning the design of computer workstations, in which the subjects' comfort votes were combined with a quantitative biomechanical analysis of postural stress to the muscles of the back, neck, shoulders and upper limbs. Independent variables were keyboard height and the presence or absence of a reading stand. The two sets of outcome measures correlated well. You would predict this, since musculo-skeletal aches and pains are commonly caused by postural stress, but it is reassuring to have it confirmed experimentally.

The type of user trial which is concerned with the layout or dimensions or a workstation or an item of furniture, is called a *fitting trial* (Jones, 1963). Figure 7.2 shows the results of such an experiment, conducted to determine the optimum height of a lectern. Ten design students (five boys and five girls) acted as subjects. A music-stand served as an adjustable mock-up. Each subject set it to the highest and lowest positions that he or she considered acceptable and also to individually preferred optimal heights. The data were fitted with normal distributions and percentages were calculated at 50 mm height intervals (Pheasant, 1984). An ergonomist would probably refer to this as a 'quick and dirty' experiment: it has certain obvious limitations, such as the extent to which design students form a representative sample of users. If the problem were more critical it might merit a more detailed investigation, but in many cases a relatively crude experiment will suffice.

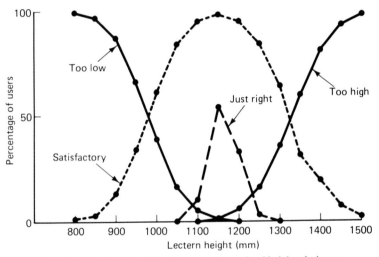

Figure 7.2 Results of a fitting trial to determine the optimal height of a lectern. (Data from Pheasant, 1984)

It can be argued that the user trial is an essential part of the iterative design cycle. I sometimes say to design students that every successful project begins with a task analysis and ends with a user trial. They often think that a project starts with a layout pad and a box of markers. (A task analysis is an operational description of what the user is intending to do with the product concerned.) In fact, the user trial may perfectly well precede the design phase as a means of gathering data; or follow it as a means of checking results.

Must every project involve experimentation or are there alternatives which the designer can use at his drawing board or computer-aided design (CAD) terminal? Many ergonomists believe the former. Over the last four decades, a sizeable collection of data has been assembled concerning the bodily dimensions and other characteristics of various user populations. It is perfectly proper, given the right circumstances, to base decisions on these anthropometric data and to avoid the hassle of running experiments. The ergonomics literature also contains accounts of solutions which have been successful in the past, or which are generally recognized as constituting good design practice in certain respects. The science of ergonomics should, in general, be moving away from evaluation and towards prediction – otherwise we are condemned to a perpetual process of reinventing the wheel. (We shall return to this matter of design standards in due course.)

7.2.3 The method of limits

I have discussed the applications of anthropometrics to design at some length elsewhere (Pheasant, 1984; 1986). They generally involve a technique known as the *method of limits*. This requires the designer to identify the category of user who will be most difficult to accommodate, and to base the design upon the dimensions of this hypothetical *limiting user*. In matters of clearance or access, the design should be based upon the bulkiest person; in matters of reach, the shortest; in strength, the weakest, and so on. In practice, the precise range of users we decide to accommodate will generally be an arbitrary judgement or be determined by other considerations or constraints. But it is often appropriate to design for a limiting user who is the 5th or 95th percentile in the relevant respect. By definition this will accommodate the 95% of the population who are less demanding in their requirements.

The concept of the limiting user has a general applicability. When the demands of the situation exceed the user's capacities to deal with them, a *critical mismatch* may be said to occur. This is what happened to the young man who lost his wages in the cardcash machine and complained that it was not foolproof. Human beings are by their very nature forgetful and inattentive; they are frequently ignorant or downright foolish. This is why the workings of things like cardcash machines must be made as clear as possible. The greatest barrier to achieving this is the unspoken assumption that other people are very much like ourselves, or that if they are not we can, by an act of imagination, place ourselves in their shoes. This belief in 'intuitive empathy' is for the most part an illusion. We simply cannot imagine what it feels like to be different from ourselves; the failure to do so results in foul-ups ranging from software to tower blocks.

Some misconceptions which can get in the way of good functional designs are given in Table 7.2. I call them the five fundamental fallacies (Pheasant, 1986). With the possible exception of number four they all pertain to human diversity. This is a

Table 7.2 The five fundamental fallacies. (From Pheasant, 1986)

Number 1	This design is satisfactory for me – it will therefore be satisfactory for everybody else
Number 2	This design is satisfactory for the average person – it will therefore be satisfactory for everybody else.
Number 3	The variability of people is so great that they cannot possibly be catered for in any design – but since people are wonderfully adaptable it doesn't matter anyway
Number 4	Ergonomics is expensive and since products are actually purchased on appearance and styling, ergonomic considerations may be conveniently ignored
Number 5	Ergonomics is an excellent idea. I always design things with ergonomics in mind, but I do it intuitively and rely on my common sense so I don't need experimental studies or tables of data

very important issue in ergonomics. Sex differences in strength, for example, pose some interesting ergonomic questions. We shall return to these shortly – but first, a brief digression.

7.2.4 Vernacular ergonomics

The tools used by craftsmen often have long and fascinating histories. In the London Science Museum (in an obscure upper gallery away from the working models which delight small boys of all ages) you will find an Ancient Egyptian stonemason's mallet, discovered near the pyramids, which is not significantly different from one which would be used today. Old craft tools are nowadays considered to be highly collectable objects; many are examples of vernacular design (and vernacular ergonomics) at its very best. They acquired their characteristic forms by an organic evolutionary process, in the context of a living craft tradition in which the tool-maker and tool-user were in constant close contact (unlike designer and user today). They exhibit, therefore, a fitness for purpose which the designer or ergonomist would find it difficult to equal. Consider, for example, the carpenter's saw. Its handle makes an angle of 70 to 80° with the cutting edge, depending upon the type. Since this keeps the user's wrist in a neutral position, as shown in Figure 7.3, it seems unlikely that this is a coincidence. Figure 7.4 shows the results of an experiment to test this hypothesis. A number of students of furniture design were asked to cut-up pieces of dowelling, as fast as they were able, using an adjustable saw. The angle found in the traditional design maximizes working efficiency as measured by speed of cutting. It also probably reduces the risk of repetitive strain injuries, which are generally believed to increase with wrist deviation.

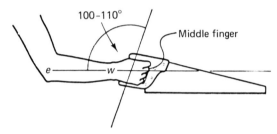

Figure 7.3 The neutral position of the wrist is preserved if the axis of grip makes an angle of 100 to 110° with the axis of the forearm (Pheasant, 1986). Reproduced by permission

Widule, Foley and Demo (1978) analysed the ergonomic history of the felling axe. The variety of axe which is most commonly used for tree-felling today became popular in North America during the late eighteenth century. It is known as the 'American axe' or 'Kentucky axe'. In his definitive historical dictionary of tools, Salaman (1975) considers this to be a misnomer, since similar axes were being manufactured in Sheffield at the same time. The axe in question had a heavier head than many of its predecessors, due principally to a very characteristic 'poll' which protruded behind its blade. Widule, Foley and Demo proposed that, as this axe had emerged as the tool of choice in a period of intensive use, then it must have offered some functional advantages. They were able to demonstrate that the heavier axe-head gave greater kinetic energy in the swing, at least in the hands of the one experienced male user they tested.

Trades involving the working of hard unyielding materials, such as wood, metal and stone, are traditionally practised by men, presumably because a fair degree of physical strength is

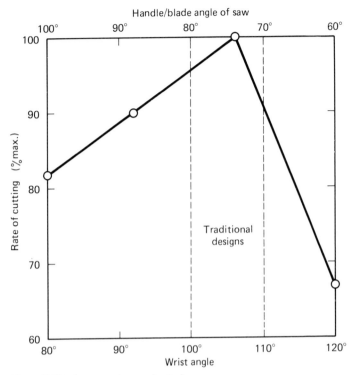

Figure 7.4 Performance in a sawing task

demanded, but women are taking-up these trades more frequently, both professionally and in home improvements. These tools presumably evolved to match the capacities of (strong) male users. Are they equally suitable for women?

7.2.5 Sex differences

Ducharme (1975) conducted a survey of female maintenance personnel in the US Air Force and found that they were frequently dissatisfied with the standard tool kits provided. Pivoting tools operated by a squeezing action such as pliers, cutters and crimpers posed particular problems. It could be argued that the handle separation of these tools, although apparently satisfactory for men, is too great for the smaller female hand. Reasonable as it is, this argument is not entirely correct. The optimal handle separation for a tool of this kind is 45 to 55 mm for both men and women as shown in Figure 7.5 (Pheasant and Scriven, 1983).

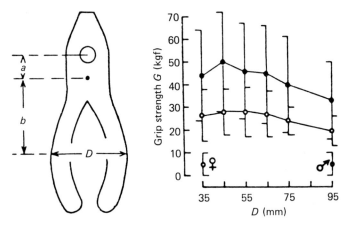

Figure 7.5 Grip strength G as a function of the handle size D. (From: Pheasant, 1986. Data from Pheasant and Scriven, 1983)

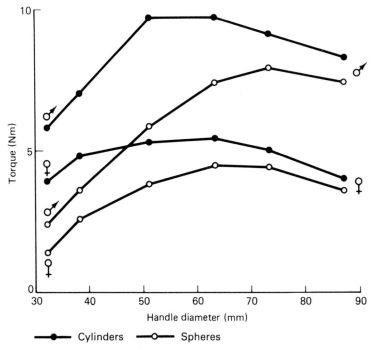

Figure 7.6 Strength of clockwise turning actions using cylindrical and spherical handles (30 male, 30 female subjects)

Although, on average, women do have smaller hands than men, the difference does not appear to be sufficiently great to be of practical consequence in the present context. The problem with these tools is actually that their mechanical advantage (usually about 6:1) is not sufficiently great for the female user to meet the relevant task demands.

We discovered something rather similar for screwdrivers. The optimum handle diameter for torque exertion is 50 to 65 mm for cylindrical handles and 65 to 75 mm for spheres, as shown in Figure 7.6 (Pheasant and O'Neill, 1975; Pheasant, 1986). Most screwdrivers on the market have an effective handle diameter of less than about 35 mm, and perform no better than a cylindrical handle of the same size (provided the surface quality of the latter is appropriate). Rather than requiring smaller handles for smaller hands, as common sense would possibly suggest, the weaker members of the population (be they women or men) would benefit from the additional mechanical advantage of a larger screwdriver handle than those readily available. This would not necessarily cause the stronger male user to wreck the heads of screws, because this is generally due to a screwdriver blade which does not fit the slot of the screw, rather than excessive torque as such.

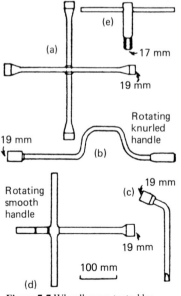

Figure 7.7 Wheelbraces tested by Pheasant and Scriven (1983)

Table 7.3 Strengths of turning actions using different designs of wheelbrace, as shown in Figure 7.7

	Men		Women	
	Mean (Nm)	SD (Nm)	Mean (Nm)	SD (Nm)
(a)				
clockwise	146	48	91	34
anti	146	42	95	42
(b)				
clockwise	51	17	35	11
anti	48	17	30	10
(c)				
clockwise	91	26	59	13
anti	95	32	59	20
(d)				
clockwise	81	32	51	15
anti	78	32	47	13
(e)				
clockwise	88	28	57	17
anti	93	29	61	20

During the course of this project, we also investigated the five motor car wheelbraces shown in Figure 7.7, all of which are supplied in the tool-kits of common makes of car, or readily available from motor accessory shops. The maximum torques which our subjects (22 men and 22 women aged between 20 and 44 years) were able to exert are also shown in Figure 7.7. A survey was also conducted of the torques which were required to undo the wheelnuts of a number of cars. A figure of 110 Nm could be regarded as typical. With the best brace (a) 20% of men and 71% of women would experience a critical mismatch, i.e. they would be unable to undo the wheelnuts without jumping on the wheelbrace, bashing it with a brick or any of the other drastic expedients we resort to when stranded at the roadside. With the worst brace (b), virtually nobody would be able to exert the necessary torque. With brace (e), supplied with a common type of British motor car, 82% of men and 99% of women were mismatched. Using these data, it is possible to determine what sort of wheelbrace would be required to enable at least 95% of women, and all men, to reach the necessary torque. A simple L-shaped box spanner, such as (c), with an effective leverage in the order of 500 mm would do the job well.

It would seem that women do not require tools that are different from those used by men. It is more that men, at least some of them, have reserves of strength that enable them to overcome deficiencies in tool design. Why are wheelbraces so unsatisfactory for such a large segment of the potential population of users? Perhaps their designers believe in the first fundamental fallacy.

7.2.6 Age differences

It is something of a cliché to say that we live in an ageing society. The number of people in the UK over 65 years old has increased by more than 2 million since 1961. Trends in birth and death rates enable us to predict future changes in the age structure of the population. Although these will not be as rapid as before, it is anticipated that the over-75 age-group in particular will increase substantially in numbers by the end of the century (Figure 7.8).

Our physical capacities decline as we get older, and we experience critical mismatches increasingly frequently. Weakness of the quadriceps muscles (at the front of the thighs) may make it difficult to get up from a low deep armchair, but the problem will be less if the chair is a little higher (within reason), more upright, has a clear space beneath the front of the seat for better foot placement and arms which can be securely gripped. A pronounced

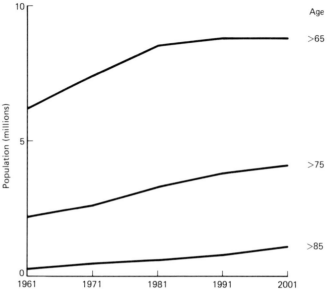

Figure 7.8 'An ageing population'. Elderly people in the UK from 1961 to 2001. (From: *Social Trends, 1985* Central Statistical Office, HMSO)

wasting of the small muscles of the hand is also common in the elderly. (The reason is obscure, but it may be due to pressure on the nerve which supplies these muscles where it passes over the first rib.) The problem may be exacerbated by stiffness and deformity of the fingers caused by arthritic diseases; or by the after-effects of earlier injuries (such as the extremely common Colles' fracture of the radius bone in the lower forearm). The result will be a pronounced difficulty in manipulating small objects such as cutlery, keys, doorknobs, etc. These problems may be greatly reduced by designing cutlery with fatter handles or fitting levers, rather than knobs, on doors.

7.2.7 People with disabilities

We may regard human abilities generally as being located in a continuum. At the one extreme are the severely 'disabled'; at the opposite are athletes, virtuosi and others with the capacity for excellence in some respect or another. In the middle are people whose capacities mainly live up to their expectations (or is it the other way around?) – for want of a better word we might be tempted to call them 'normal'. This is a rather deceptive word. In the present context it means something like 'averagely unfit'. The boundaries between normality and disability are fuzzy and arbitrary. At best they are relative concepts. To suffer a disability means to lack an ability we would like to have, presumably in order to do something we wish to do, and is therefore to experience a critical mismatch in a certain range of tasks. (This makes us all disabled.) But the difficulties experienced by the person with disabilities are to some extent design-related. The wheelchair user is confronted by a variety of obstacles to his mobility such as kerbstones, stairs, revolving doors, lavatories without adequate turning space, and so on. All of these have been put there by other people.

The ergonomic issues posed by people with severe disabilities are somewhat specialized. But in the grey area between disability and normality are a very large number of people who are little more than averagely unfit. (Their condition may only be temporary – a sprained ankle, a touch of 'flu, lack of sleep, etc.) These *less able users* wish to undertake a full range of everyday activities, but they experience a more-than-usual degree of pain or difficulty in doing so. This makes them particularly intolerant of bad design.

The standard height for a kitchen worktop, as specified in BS 6222 and ISO 3055, is 900 mm. This is satisfactory for the shorter 65% of women or 45% of men (Pheasant, 1986). Taller people will generally find it a little too low. A kitchen sink set directly into the worktop (as is almost always the case) is lower than desirable for most people. Bathroom handbasins which are generally less than 800 mm in height are only really suitable for children and, perhaps, wheelchair-users (Pheasant, 1986; Kira, 1976). A reasonably fit person will tolerate a modest deviation from the optimum height, provided he does not have to tolerate it for too long. But the tall backpain sufferer, of whom there are many, will find the task of washing up at a standard sink an unbearably painful experience.

It is not uncommon to find that ergonomic improvements, made for the benefit of people with disabilities, would also make the products easier and more comfortable to use by the able-bodied; ergonomic niceties, which the average person may look upon as inessential, may transform the life of the less-able user, e.g. the severe neck and shoulder problems of a middle-aged typist have improved overnight when she was provided with a suitable raised reading-stand.

A cautionary tale is in order at this point. Some years ago I was holding forth to an architect friend of mine about the desirability of having kitchen worktops in more than one height. He replied that if he had to worry about things like that he would never get a building finished at all. More recently, a colleague and I wrote an article in the *Architects' Journal*, in which we suggested that a single uniform height for all kitchen worktops was not an ideal arrangement (Drake and Pheasant, 1984). It raised howls of protest in the letters column, the essence of which was that such an arrangement was unrealistic.

7.2.8 Standards

I am therefore increasingly convinced of the importance of standards. Designers require well-defined units of ergonomics information for incorporation into their products – rather like assembling a set of standardized modular components into a larger structure. This is not to denigrate designers in any way whatever; as my architect friend forcefully pointed out, they simply have other things to think about. That is why I recently spent a year or so compiling a 'cookbook' of all the current British and international standards which I could identify as having a recognizable ergonomic content, together with a collection of other relevant data and guidelines gathered from here and there (Pheasant, 1987).

Some of these standards are based on substantial bodies of ergonomic research and may be considered definitive. This category would include, for example, BS 3693, which deals with analogue displays, and ISO 7001 (with ISO/TR 7239) which deal with public information symbols. Others have not been established in quite this way, but nonetheless represent a consensus of what constitutes good design practice, as currently conceived, e.g. BS 4099 which deals with indicator lights and includes a set of standardized meanings for their colours, which are essentially an attempt to codify users' expectations.

I was struck, while compiling these guidelines, by what William Blake, in his prologue to the poem *Jerusalem*, wrote concerning those who would do good to others, namely that they must '. . . do it in/Minute Particulars'. User-centred design is very much a matter of dealing with the Minute Particulars. This is what some people call the 'bottom-up' approach to design, in contrast to 'top-down' which attempts to derive the details of a design from some overall grand theory. It was grand theories, concerning what people ought to require, that brought us the tower block, the open-plan office and the windowless schoolroom (which was supposed to help children concentrate by cutting off outside distractions). It was a grand theory (predominantly in the area of aesthetics) which led to all those products of the school of design known as functionalism, which were characterized by their lack of ornament and brutality of appearance rather than any real consideration of fitness for human purpose.

The bottom line may be that the best we can hope for is that products will conform to the available ergonomic standards and guidelines. In pragmatic terms this may well be what the term 'ergonomically designed' will often have to mean. If this alone were achieved, our man-made environment would be a more comfortable and less confusing place (and no less interesting for so being), and the catastrophes we considered at the outset would occur a good deal less frequently. We also need some common sense. But in this imperfect world, common sense is a commodity which is in short supply.

7.3 Postscript

As I was assembling the typescript of this chapter, I noticed the following sign taped to the top of the photocopier: Before using this machine for the first time, please consult the library staff. Enough said?

References

BARTRAM, D., CRAWSHAW, C. M. and WILLIAMS, D. I. (1985) *The use of heating controls*. Ergonomics Research Group Report No. ERG/Y6543/85/1. ERG, University of Hull.

CONRAD, R. (1962) 'The design of information'. *Occupat. Psych.*, **36**, 159–162

DALE, H. C. A. and CRAWSHAW, C. M. (1983) 'Ergonomic aspects of heating controls'. *Building Serv. Engng Res. & Technol.*, **4**, 22–25

DRAKE, F. and PHEASANT, S. T. (1984) 'Domestic kitchen design – the ergonomist's view'. *Arch. J.*, 3 Oct., 79–80

DUCHARME, R. E. (1975) 'Problem tools for women'. *Ind. Engng*, **7**, 46–50

JONES, J. C. (1963) 'Fitting trials – a method of fitting equipment dimensions to variations in the activities, comfort requirements and body sizes of users'. *Arch. J.*, 6 Feb., 321–325

KIRA, A. (1976) *The bathroom*. Penguin, London

LIFE, M. A. and PHEASANT, S. T. (1984) 'An integrated approach to the study of posture in keyboard operation'. *Appl. Ergonomics*, **15**, 83–90

PHEASANT, S. T. (1984) *Anthropometrics – an introduction for schools and colleges*. PP7310. British Standards Insitution, Milton Keynes

PHEASANT, S. T. (1986) *Bodyspace – anthropometry, ergonomics and design*. Taylor and Francis, London

PHEASANT, S. T. (1987) *Ergonomics – standards and guidelines for designers*. PP7317. British Standards Institution, Milton Keynes

PHEASANT, S. T. (1988) 'Fatal errors – The Zeebrugge/Harrisburg syndrome', *New Scientist*, 21 Jan. 55–58

PHEASANT, S. T. and O'NEILL (1975) 'Performance in gripping and turning – a study in hand–handle interaction'. *Appl. Ergonomics*, **6**, 205–208

PHEASANT, S. T. and SCRIVEN, J. G. (1983) 'Sex differences in strength – some implications for hand-tool design'. In: K. Coombes (ed.) *Proceedings, Ergonomic Society's Conference*. Taylor and Francis, London, pp. 9–13

SALAMAN, R. A. (1975) *Dictionary of tools used in the woodworking and allied trades, c. 1700–1790*. Allen and Unwin, London

WIDULE, C. J., FOLEY, V. and DEMO, G. (1978) 'Dynamics of the axe swing'. *Ergonomics*, **21**, 925–930

WILLIAMS, D. and CRAWSHAW, C. (1986) 'Advanced control systems and the householder'. In: D. J. Oborne (ed.) *Contemporary ergonomics 1986*. Taylor and Francis, London, pp. 325–329

Chapter 8

The ergonomics of protection

N. T. Thomas

8.1 Introduction

Human beings have always needed to be protected against the environment. Clothing, like food and water, is a fundamental human need. In this sophisticated age of relatively high living standards and consumer spending in the developed countries, much time and attention is expended on the clothes we select and wear. The industry employed on designing, manufacturing, testing, cleaning and criticizing clothing has grown hand in hand with this demand, responding to the enormous demands by technological innovation.

Two parallel developments have taken place in recent times: (1) an increasing proportion of the population has been able to choose a range of garments for different functions, e.g. workwear, leisure and sportswear and casual clothing; and (2) designs and materials for protective clothing have improved to cope with more sophisticated environments and hazards. Interaction between these changes can have significant effects on whether the wearer is safe and remains healthy in both the short and long term, i.e. the design of clothing and material needs to match the task for which it is intended. Thus, it is vitally important that the ergonomics of protection needs to be examined, understood and applied for workers in industry today, so that exposure to countless hazards can be controlled.

Ergonomics design aims to match the capabilities of workers to the work they are doing: in the context of protection this may take two forms:

1. In its broadest sense, there is the long-term need, if feasible, to isolate or contain the source of hazards. Control of exposure should be secured by measures other than the provision of personal protective equipment (PPE) if reasonably practicable (Health and Safety Executive, 1987). The decision to require the wearing of protection should not be the only option considered, nor the first priority, except in cases of emergency.

2. The measures taken to prevent, or provide control of, exposure to hazards are often insufficient and unsuitable. Then the design of necessary protective clothing and equipment has to be optimized in relation both to the required function and to the working wearers – their capabilities, limitations and effectiveness.

In assessing the ergonomic needs, it is important to establish the objectives and the approach to be adopted before considering specific requirements.

8.2 Objectives and approach

Objectives and approach are best considered in terms of three main factors: (1) long-term objective; (2) short-term objective; and (3) approach.

1. *Long-term objective.* Workers really *must have* ergonomic protection. The whole working system has to function effectively, possibly without relying upon PPE.
2. *Short-term objective.* Illustrates *how* workers may achieve this effective protection by exploiting ergonomic criteria. Here the exploitation is of available knowledge, not of available workers.
3. *Approach.* A traditional engineering approach to protection concentrates upon engineering control *before* PPE. This approach is also sound ergonomics as it generally obviates the need to consider the variability of people and their behaviour. Such variation across the range of human factors can create many problems which may be difficult to control. The first approach to be considered, therefore, is one based on engineering control; although this approach may require a greater capital investment and development time, the longer-term advantages can be very attractive, as discussed in other chapters.

   There are situations, however, where protection is the most effective and expedient method of control. Thus, the second ergonomic approach is to optimize the fit between the protective equipment and the worker. This chapter will focus upon the implications of this interaction.

   It is first of all important to note that the title of this chapter is 'The ergonomics of protection' without the addition of 'equipment'. This is deliberate, as the ergonomic approach should not concentrate solely upon the equipment nor upon the techniques of ergonomics, but upon the wearer. A priority sequence is needed as below.

(a) *emphasize the people,* who need to be protected, particularly their characteristics. These range from 'strengths', such as the ability to cope with discomfort, to 'weaknesses', such as changes in behaviour, like taking risks as the time of exposure to a hazard is increased.

(b) *consider the hazard itself* and how it is perceived by those exposed to it. A worker's perception of a hazard possibly differs greatly from that of the manager or engineer responsible for its regulation of initial design and installation.

(c) *Examine the available options and consequences* in both the long and short term. It is stressed that the effectiveness of PPE is strongly dependent upon its *acceptability*, and hence the actual exposure time when the equipment is worn properly.

This priority sequence is illustrated in Figure 8.1. Acceptability depends upon:

(a) Comfort, or more probably discomfort from the protection;

(b) Ease of use; and

(c) Perceived protection.

These variables are discussed in the following sections.

Figure 8.1 The priority sequence for considering the ergonomics of protection

8.3 Perception of acceptability

To what extent is a wearer able to assess accurately the risks of wearing or not wearing protection? The most relevant criterion is the way in which discomfort is tolerated. Because of this coping mechanism a worker deals in sequence with three stages as hazards are confronted: (1) the hazards; (2) the related discomfort; and (3) the working time.

1. *The worker's appreciation of the hazard's relative magnitude.* It is vitally important that, in order to assess the risk from exposing himself to the hazard, knowledge of the hazards are communicated in a meaningful form so that awareness of the need for protection becomes clear. Despite the Health and Safety at Work, etc. Act 1974 such knowledge seems to be regularly overlooked in industry, and is therefore considered in greater detail in this chapter.

2. *Protective equipment is likely to be an intrusion on personal comfort* by the very nature of its function, and some equipment is more likely to lead to a feeling of discomfort than others. The ability to cope with discomfort is the second stage, and has a great bearing on the continued use of the equipment. If the discomfort exceeds a subjectively determined threshold, even one which has been formulated with a knowledge of the risks he or she will be running by working unprotected, it will be much more likely that protection will be discarded. Crockford (1981) has stated that such equipment would be worn for protection only if the discomfort and danger from the equipment are perceived to be less than those from the hazard itself.

    Figure 8.2 serves to illustrate the many operational factors which influence the acceptability of a respirator system.

3. *The time of working while protected.* There is a limiting time for wearing any type of clothing or equipment. In addition, as familiarity to the working conditions and hazard increases, so 'contempt breeds'. This probably leads to a decreased perception of danger and, hence, a greater reluctance towards effective protection. Both of these tendencies focus attention upon the need:

    (a) to select target working times for protected working (Thomas, 1977); and

    (b) to reinforce an enhanced perception of danger at regular intervals, so that protective equipment may be worn effectively. Behavioural technology (considered later in this chapter) is one technique used for the reinforcement of a danger.

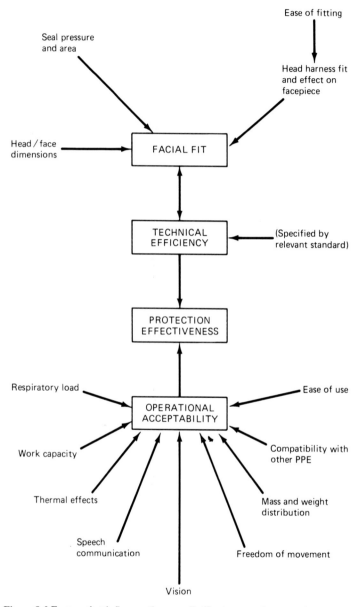

Figure 8.2 Factors that influence the overall effectiveness of any respirator system.
From: Graves, (1985) (Figure 3)

Having summarized the ergonomic approach to protection, it is appropriate to examine how ergonomic knowledge can be applied to the design and selection of personal protective equipment.

8.4 Design and selection of personal protective equipment

Most protection problems are best solved by a combination of three basic applied sciences: (1) anatomy; (2) physiology; and (3) psychology.

8.4.1 Anatomy and anthropometry

Anatomy has an obvious relevance to the design of clothing and equipment linked through anthropometry – the systematic measurement of human dimensions. It is important to cater for the range of equipment users and to consider both the upper and lower extremes (the 5th and 95th percentiles) of the dimensions relevant to the equipment to be worn. A concentration upon the average person must be avoided. Even if only three average body dimensions such as stature, arm length and head circumference are considered, then it is difficult to find one person with the three corresponding dimensions. The size and shape of the relevant equipment available must cater for all the workers who are actually exposed to the hazard.

New problems can soon arise although, at first sight, a cost-cutting exercise may appear to warrant a reduction in available sizes. Hidden costs and threats to safety grow in proportion to discomfort. It has been found, for instance, that glove design can reduce finger mobility, and therefore productivity, in both thermally neutral and cold environments (Parsons and Egerton, 1985). In another study, it was found that wearing gloves can increase task times by up to 50% (Thomas, Spencer and Davies, 1976). The design of one type of ear-muff is only effective for a range of head size up to the average (mean) size plus *one* standard deviation. This means that the design is unacceptable for about one-sixth of the working population.

Much industrial clothing and equipment is made in too few sizes and to out-of-date anthropometric data (or, more probably, traditional 'rule of thumb') and with excessive girth allowances (Bolton, 1975). Work uniform, shoes and goggles have also been identified (Ducharme, 1977) as inadequate in more than one craft skill by between 12 and 100% of 1400 US Air Force women working in the different craft skills. As changes in standards occur (e.g. British Standards) so new opportunities arise to classify clothing into new sizes, appropriate to the latest anthropometric data available.

8.4.2 Physiology

Physiology has particular relevance to restrictions in posture and heat dissipation. The heat balance of the body is controlled via the mechanisms of convection, conduction and radiation from the skin. Clothing creates a 'micro-climate' between the clothing and the skin; where the clothing is impermeable to air, the major mechanisms of heat loss, evaporation and convection are affected adversely and heat is retained in the micro-climate thus reducing heat loss from the body. The action of the muscles during work produces heat which is transferred via the bloodstream to the skin. If the clothing is inappropriate, the heat is prevented from escaping beyond the clothing barrier and body heat and fatigue quickly build-up. Details of heat exchange while wearing protective clothing have been explained and reviewed in a paper by Parsons (1988). Relevant terms have been revised by the International Union of Physiological Sciences Thermal Commission (1987). Some of these terms are included below.

Increases or decreases in the body heat content are limited.

The positive or negative storage of heat (S) in the body is expressed in the body heat balance equation (International Union of Physiological Sciences Thermal Commission, 1987) which eventually gives $S=0$ as the body heat content is stabilized:

$$S = M-(W)-(E)-(C)-(K)-(R)$$

where M is the metabolic energy transformation (always positive), W is the work (positive if useful work is accomplished, negative if mechanical work is absorbed by the body), E is the evaporative heat transfer (usually positive if heat lost), C is the convective heat transfer (can be positive or negative), K is the conductive heat transfer (usually this term can be ignored, except in local areas or if the body is immersed in liquid), and R is the radiant heat exchange (can be positive or negative).

The *dry* heat loss is mainly given by $C+R$.

Clothing and equipment have a thermal insulation value against this dry heat expressed in 'clo' units where $1\,\text{clo} = 0.18°C \cdot M^2 \cdot \text{h} \cdot \text{kcal}^{-1} = 0.155°C \cdot M^2 \cdot W^{-1}$. (These units are based on the Systeme Internationale (SI).) A person whose clothing has an insulative value of 1 clo when sitting at rest in 'comfortable' indoor surroundings (21°C) has a metabolic rate of 1 'met', where 1 met = $58.15 W \cdot m^{-2} = 50\, kcal \cdot h^{-1} \cdot m^{-2}$. 'Thermal comfort' is subjective indifference to the thermal environment.

The effective insulation of clothing is $I_{cl} + I_a$, where I_a is the reciprocal of the thermal conductance of the ambient environment. $I_{cl} + I_a$ is usually measured as the temperature gradient from the surface of a heated man-sized manikin to the ambient air divided by the heat production per unit area of manikin surface. The units are $°C \cdot m^2 \cdot W^{-1}$, sometimes expressed in clo units.

Physiological studies can produce clear objective data, and therefore much protection has been ergonomically improved. Both fabric materials and the construction of garments have benefited from applied physiology.

8.4.3 Psychology

Psychology is the most recent basic science to have been applied in ergonomics and has much to offer in the appropriate selection and use of protective equipment. Although the data obtained from applied psychology is often less clear than that from applied physiology or anatomy, subjective and behavioural data can be easier to gather and possibly be more directly applied. For example, a scientific laboratory evaluation of clothing which produces adverse subjective reactions is likely to correlate with similar adverse responses in field behaviour, such as an unwillingness to wear ill-fitting equipment.

The various ways in which groups and individuals actually *select* their protection for different tasks, against different hazards, has a significant effect upon subsequent behaviour. The initial choice may be made by management, a safety committee or by the wearers themselves. Each group may have particular interests in selecting certain types of equipment and there is a need for a detailed assessment of the requirements to cover comprehensively all the many considerations which are relevant. These may include the contract of employment, education and training, function, fitness, purchasing schedule, testing, storage and shelf-life, and inspection and cleaning. Laundering may cause a considerable decrease in protective functions leading to the need for more frequent repair, maintenance and, ultimately, replacement. A further factor is the quality of supervision – it may be necessary to educate supervisory staff of equipment use, function and availability as a means of reinforcing its use by the workers.

Effective use of PPE is relevant in particular to two applications of psychology: (1) the signal detection theory; and (2) behavioural technology.

8.4.3.1 Signal detection theory (SDT)

Transmitted hazard information may be interpreted as 'signals'. If these signals, or the probability of their presence, are sensed above the background information, or 'noise', and are accepted by the worker, it may be assumed that protection would be used.

Signal detection theory is explained in a readable introduction by one of its exponents (Swets, 1973). The basis of the theory is that the relationship between sensory effect and the probability of its detection varies according to the left-hand distribution in Figure 8.3 when S_0, noise, alone is present, and according to the right-hand distribution, S_1, when a given signal is added to the noise. The criterion for a positive response, c, is assumed to be fixed at such a point that it is rarely exceeded by noise alone, with no discrimination possible below that point. Positive responses below c, therefore, can be considered to be random guesses.

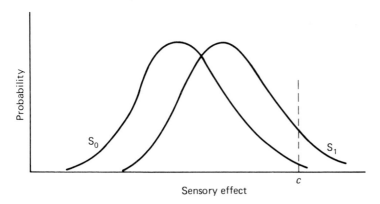

Figure 8.3 The basis of the signal detection theory. (From Swets (1973))

Thus, SDT attempts to distinguish between the worker's *ability* to detect such signals and the *probability* that he or she will react correctly to them. 'Correct reaction', from a health and safety viewpoint, may be assumed to take the form of continuous use of protection where there is a possibility that the hazards may be present. Applying a modified form of SDT, such a continuous use would occur when the wearer of protection decides that he or she *may* have perceived a signal, even if it were only a 'false alarm'. A missed signal is undesirable as it could lead to an accident or ill health; consequently, 'false alarms' are tolerated in the process towards improved health and safety at work.

There are at least four lessons to be learned from the SDT for ergonomic protection: (1) sensing the hazard; (2) previous information; (3) the signal itself; and (4) the worker's motivation.

1. *Sensing the hazard signal.* The most obvious aspect is that the information from the hazard should be sensed in some form. It should be seen, heard or smelt, for example, above the background information or 'noise'. This draws attention to the need for a boost to the signal, such as the addition of a colour, or 'smell agent' or, alternatively, to the need to remove 'noise' from the hazardous workplace. The design of protective equipment and clothing often has a significant impact upon the 'noise' transmitted to the wearer (see 'Noise' and a false sense of security, below).

2. *Previous information.* If clear and striking information on the hazard and its effects is provided in advance, protection against the hazard is more likely to be effective. A useful example of such information feedback to encourage regular use of hearing protectors was applied by Zohar, Cohen and Azar (1980).

3. *The information content of the hazard signal is also important to the receiver.* Some signals, such as strong unpleasant odours or meaningful smoke or flames, prompt a speedy response instinctively. This natural reaction could also be investigated by following up a related ergonomic concept – the population stereotype (Murrell, 1965). Thus, by utilizing workers' known expectations and reactions when confronted with a particular situation, it may be possible to reinforce the required response. Artificial 'boosting' of a signal, as mentioned in (1) above, may also be designed in accordance with instinctive responses, e.g. to the colour red.

4. *The worker's motivation.* When the known penalty for failing to respond to a signal to use protection is severe, such as a likelihood of serious injury or an industrial disease, an effective response is probable. A similar desirable response may also be encouraged by positive external incentives from the worker's supervisors. The dangers, or lack of effectiveness, of negative incentives (Zohar, Cohen and Azar, 1980) should also be heeded. They found that 'punishment' of those ignoring the need for protection had not been effective.

'Noise' and a false sense of security   The four lessons from the SDT explained above, should also be viewed in the context of the modifications in behaviour that people make when protection is

worn in a hazardous environment. If the protective barrier does not allow its wearer to discriminate clearly between background noise and the additional hazard signal, then the wearer may miss the signal. The protection itself increases 'noise'.

In the relatively comfortable protective micro-climate, the wearer may gain a false sense of security. This could direct attention away from the dangers. The wearer may then fail to accept warning signals of the hazard. Information on safety and health protection may be missed. Such a situation has been explained (Thomas, 1977) as 'insensitive protection' whereas 'protective fatigue' may be desirable. 'Protective fatigue' is a concept whereby a level of 'fatigue' is desirable in order that the wearer may accept the signal and suitably modify his behaviour as a result.

8.4.3.2 Behavioural technology

Reduced exposure to hazards is usually recommended by engineering controls (Health and Safety Executive, 1987). Another approach, that of behavioural technology (Hopkins, *et al.*, 1986) has been demonstrated with success in the short term, and on a much larger scale over 100 work days (Hopkins, Conrad and Smith, 1986).

Behavioural technology comprises the teaching of selecting safe behaviour to workers, followed by a behaviour maintenance programme to encourage continued use after training. Hopkins, *et al.* (1986) showed a strategy for developing and testing behavioural technology. Workers were found to improve their covering of skin with gloves and sleeves, while another sprayer increased his use of a protective respirator.

The long-term reliability of this form of control needs to be examined in detail, but it has been found to become more reliable and effective over the moderate time intervals (Hopkins, Conrad and Smith, 1986). Further study of this approach is proceeding.

8.5 The way ahead

The Health and Safety at Work, etc. Act specifies that adequate information and tests, as well as necessary research on work articles, is required to eliminate or minimize risks. Research needs for improved protective clothing have been identified in the US (NIOSH 1986). International (International Organization for Standardization, 1987) and British Standards are being developed at the frontiers of clothing knowledge. New guidance from the

Health and Safety Executive on protective clothing is being prepared. Some developments at the frontiers of knowledge are possibly premature. An example is the proposed use (International Organization for Standardization, 1987) of estimated values of dry heat insulation (clo) for clothing worn in thermal environments. Scientists from various disciplines have expressed concern on the accuracy and reproducibility of such estimates. There needs to be further research on the dissipation of body heat by evaporation of sweat through clothing and protective equipment. The Consumer Credit Act in the UK and new laws in Europe on product liability may encourage manufacturers and producers to carry out the necessary research and testing. One hopes that the industry will then provide more information on their products so that the accuracy of standards can be reviewed and improved in the light of this data.

Ergonomics crosses and applies disciplines to reduce protection risks. The Ergonomics Society and the British Occupational Hygiene Society have recently set up a Clothing Science Group in the UK with approximately 100 participants. The Clothing Science Group has brought together applied scientists, clothing designers, suppliers and users. By the presentation of papers, workshop sessions and informal interchange, developments and standards in materials, design and research can be progressed. A safer and more healthy industry is one of the aims of the Group.

8.6 Conclusions

It is possible to design safer and more effective protection for workers by applying the basic sciences of ergonomics, as outlined above. There is also considerable promise with respect to the selection and appropriate use of the equipment. With information on:

1. the characteristics of people – their capabilities and limitations;
2. the hazardous environment and its perception by workers; and
3. the vital acceptability of protective equipment, work itself can be designed to be safer and more healthy. Available and developing knowledge needs to be used and exploited. There is no need for workers' health or life-styles or, indeed, their survival to be ruined just because their protection is not ergonomic.

8.7 Acknowledgements

I am indebted to the Health and Safety Executive for its guidance and for the exciting challenges of four investigations into protective clothing and equipment since 1982.

References

BOLTON, C. B. (1975) 'Design of functional clothing', *Ergonomics*, **4**, 470

CROCKFORD, G. W. (1981) 'Protective clothing and equipment', In: *Occupational health practice.* Schilling R. S. F. (ed.). Butterworth, London

DUCHARME, R. E. (1977) 'Women workers rate 'male' tools inadequate', *Human Factors Society Bulletin*, **20**, 4, 1–2

GRAVES, R. (1985) 'Personal protective equipment: the ergonomic implications', *The Safety Practitioner 3*(9), 6

HER MAJESTY'S STATIONERY OFFICE (1974) Health and Safety at Work, etc. Act. HMSO, London

HEALTH AND SAFETY EXECUTIVE (1986) *Occupational exposure limits – Guidance Note EH40/86.* HMSO, London

HEALTH AND SAFETY EXECUTIVE (1987) *The control of substances hazardous to health regulations* (July draft). HMSO, London

HOPKINS, B. L., CONRAD, R. J., DANEL, R. F., FITCH, H. G., SMITH, M. J. and ANGER, W. K. (1986) 'Behavioural technology for reducing occupational exposures to styrene'. *J. Appl. Behaviour Analysis.*, **19**, 1, 3–11

HOPKINS, B. L., CONRAD, R. J. and SMITH, M. J. (1986) 'Effective and reliable behavioural control technology'. *Am. Ind. Hygiene Assoc. J.*, **47**, 12, 785–791

INTERNATIONAL ORGANIZATION FOR STANDARDIZATION (1987) *Draft standard – estimation of the thermal characteristics of a clothing ensemble,* Publication No. ISO/TC 159/SC5N 68E. ISO, Paris

INTERNATIONAL UNION OF PHYSIOLOGICAL SCIENCES THERMAL COMMISSION (1987) 'Glossary of terms for thermal physiology'. *Pflugers Arch*, **410**, 567–587

MURRELL, K. F. H. (1965) *Ergonomics – man in his working environment.* Chapman and Hall, London

NATIONAL INSTITUTE FOR OCCUPATIONAL SAFETY AND HEALTH (1986) *Criteria for a recommended standard – occupational exposure to hot environments, revised criteria.* NIOSH, Cincinnati

PARSONS, K. C. and EGERTON, D. W. (1985) 'The effect of glove desing on manual dexterity in neutral and cold conditions'. In: D. J. Oborne, (ed.) *Contemporary ergonomics.* Taylor and Francis, London

PARSONS, K. C. (1988) 'Protective clothing: heat exchange and physiological objectives'. *Proceedings, Clothing Science Group Conference* (in press)

SWETS, J. A. (1973) 'The relative operating characteristic in psychology'. *Science*, **182**, 990–1000

THOMAS, N. T., SPENCER, J. and DAVIES, B. T. (1976) 'A comparison of reactions to industrial protective clothing. *Annals of Occup. Hygiene*, **19**, 259–268

THOMAS, N. T. (1977) 'Investigations into human performance and industrial fatigue'. M. Phil. thesis. The Polytechnic of Wales, Pontypridd

ZOHAR, D., COHEN, A. and AZAR, N. (1980) 'Promoting increased use of ear protectors in noise through information feedback'. *Human Factors*, **22**, 1, 69–79

Part 4

Perspectives on current issues

Chapter 9

Work-related musculo-skeletal disorders

P. W. Buckle

9.1 Introduction

There is an implicit message contained in the words 'work-related' which leads one to suppose that the disorders to be discussed in this chapter are brought about by work or by occupations. However, this would be a misinterpretation as the real message is in the *association* between work factors and specific musculo-skeletal disorders. The concept of causal and associated relationships will be returned to in section 9.3, although it should be said here that failure to recognize this important distinction has led to much of the confusion in the scientific literature. This has inevitably spilled over into other areas, notably that of preventive approaches.

It is evident that in order to investigate such problems, and to justify their research, we must first have a clear understanding of the patterns of these states of health. It also would be useful to know what brings these states about and, if possible, what *prevents* the development of ill health. The branch of scientific discipline most concerned with such studies is epidemiology, and it is data from such methods which occupy section 9.2.

The majority of epidemiological studies provide us with evidence for risk factors associated with the disorder(s) under investigation. They rarely generate what might be termed a cause-and-effect relationship, although they do provide extremely useful data for setting up such hypotheses and directing research into the underlying pathology of the disorders. Risk factors associated with musculo-skeletal disorders will be considered in section 9.3.

The object of the exercise is, of course, one of prevention. Prevention may take one of two forms: (1) primary, aimed at preventing the problem arising; and (2) secondary, concerned with preventing the recurrence of the problem in those already affected. In section 9.4 I have attempted to deal with approaches commonly used in preventive strategies. I have chosen to

113

concentrate specifically on primary prevention as this is more in keeping with ergonomic philosophy. Examples of how some preventive strategies have helped and how others have failed will also be discussed.

Finally, in section 9.5 I have attempted to look to the future and to consider some problems which need to be resolved if real progress is to be made in reducing the prevalence of a set of disorders which have become of such major concern both for sufferers and society at large.

9.2 Epidemiological considerations

A major difficulty exists in presenting statistics of an epidemiological nature in an interesting manner; without them, however, we have no real basis for justifying research into this area. Similarly, it makes the planning of health care difficult and the targeting of resources for investigation of disorders unsystematic. Such data can, and do, provide us with evidence of changing trends in disorders and allow us to take action before the problem gets out of hand.

Data provided by the Department of Health and Social Security, as cited by the Office of Health Economics (1985), allow us to estimate that 61.4 million working days were lost in 1982/83 as a result of disorders of the musculo-skeletal system (Table 9.1). In the current economic climate it is often necessary to consider the cost of such disorders before they receive attention. For back disorders (Office of Health Economics, 1985) the estimated cost is detailed in Table 9.2.

These costs can be put into some context when we realize that the potential savings to the NHS alone could, in theory anyway, have made possible a 20% cut in hospital waiting lists in 1986. The

Table 9.1 Days of certified incapacity for work, 1982/83.
(From Office of Health Economics (1985))

Condition	(%)
Circulatory	18
Musculo-skeletal system	17
Mental disorders	14.5
Respiratory	13.2
Accidents	9.2
(Total days 361 million)	

Table 9.2 Non-National Heatlh Service costs, 1982.
(From Office of Health Economics (1985))

33.1 million days of certified incapacity
£1 018 million of potential output lost

Cost of back pain to the National Health Service, 1982

	(£million)
General medical services	25.7
Pharmaceutical services	38.9
Out-patient consultations	25.3
Hospital in-patients	66.2
	£156.1 million

potential savings are enormous but research investment when set against these figures is barely noticeable.

I have looked elsewhere for data on other disorders of the musculo-skeletal system. The centrally generated records of prescribed industrial diseases make interesting reading. Figures 9.1 to 9.3 illustrate how tenosynovitis (inflammation of the tendon sheath) ranks as a percentage of the overall problem (Figure 9.1).

A time analysis (Figure 9.2) illustrates how a substantial reduction in the prevalence of dermatitis has occurred in the time

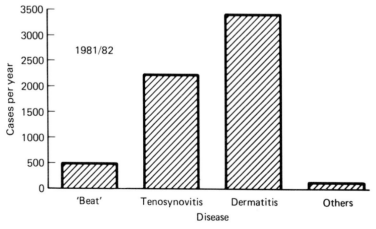

Figure 9.1 Prescribed industrial diseases, spells of benefit (Office of Health Economics)

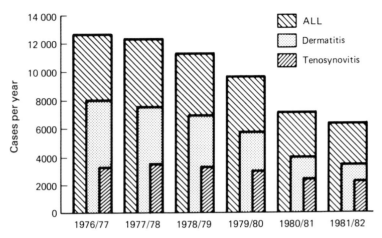

Figure 9.2 Prescribed industrial diseases from 1976 to 1982 (relationship between tenosynovitis, dermatitis and all diseases) (Office of Health Economics)

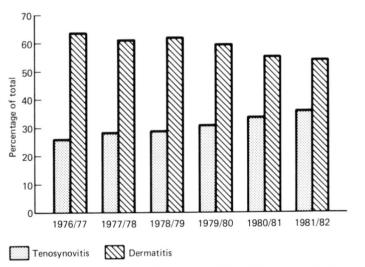

Figure 9.3 Prescribed industrial diseases from 1976 to 1982; (relationship between tenosynovitis and dermatitis) (Office of Health Economics)

span studied, but that only a small reduction has been observed for tenosynovitis.

Figure 9.3 shows that the relative significance of tenosynovitis as a percentage of all prescribed industrial diseases has therefore increased substantially. Before being accused of using and abusing the statistical data too much and while not wanting to say too much about their quality, I will move on to the next section. (For those who are addicted to such data sets I would suggest Jayson (1987) and Kelsey (1982) for further reading.)

9.3 Causes and associations

It was mentioned in section 9.1 that causal relationships are few and far between in this area. Where they do exist (e.g. after gross mechanical trauma) the underlying pain pathology may still be confused. It is usually better, therefore, as a general rule to consider most of the findings of such epidemiological studies in the context of associated risk factors and to await advances in research on the underlying pathology before making statements on causal relationships. Good arguments along such lines for disorders of the upper extremities have been advanced, amongst others, by Hadler (1985).

Turning, then, to what is known of the associated risk factors, we rapidly become submerged in a sea of diverse and often conflicting results. Problems of epidemiological study design, especially with respect to exposure to hypothesized risk factors and in developing suitable diagnostic criteria for recognition of cases (or sufferers), prove a perpetual headache to any reviewer. Those who have attempted this in recent years and who have provided a more detailed account than space allows here include Troup and Edwards (1985) and Wallace and Buckle (1987).

Perhaps the only saving grace of these methodological complexities is that for some disorders, notably back pain, there have been many researchers who have 'had a go'. The majority of these investigations are of a retrospective or cross-sectional nature. The findings of such studies have identified a multifactorial web of associated risk factors with only a few studies attempting to combine these in an acceptable statistical manner, e.g. that by Punnett, Fine and Keyserling (1987).

It is evident that greater effort needs to be put into ascertaining the degree of risk associated with such conditions; similarly, work must be done on better quantification of the exposure to tasks. This must encompass the biomechanical forces involved, the degree of movement and the repetitive parameters of such tasks. It is only through the collection of such data that effective guidelines to exposure based on health-dependent variables can be generated. One particular area of omission in the existing literature is the frequent failure to indicate to what extent static muscle loading is a factor in the development of such disorders. There is increasing evidence that the effects of static loading are characterized by the rapid onset of localized fatigue and of chronic discomfort. Far too much emphasis is still placed on the parts of the body which are moving during a work task. The relationship between static and dynamic muscular work is under-researched.

There is a similar paucity of data relating to, for example, the forces which working populations can exert. Those data that do exist have usually been collected from highly unrepresentative groups (e.g. US Air Force personnel) and are of little or no use in attempting to provide designers of, for example, hand-held tools, with guidelines for the strength capabilities of the female workforce in the UK.

9.4 Approaches to prevention

Snook, Campanelli and Hart (1978) advanced three common approaches to prevention of back-pain problems in industry. While this chapter deals with a wider range of disorders, it seems that the approaches put forward by Snook, Campanelli and Hart would apply equally well to most of them. The approaches are (1) training; (2) selection and screening; and (3) ergonomics. I shall deal with each in turn.

9.4.1 Training

There is little or no evidence that training in safe handling techniques has had any real impact on the reduction of back-pain within industrial workforces. The reasons for this may be partly methodological in that there are considerable difficulties in undertaking good epidemiological studies to test the efficacy of such an approach, or perhaps there are similar problems relating to the ability to train people in what might be complex skills requiring high degrees of psychomotor co-ordination. The training is all too often inappropriate because it cannot be applied within the constraints of the real work situation. The need to remove these constraints would seem to be a prerequisite and will be considered further below. Furthermore, the extent to which training in such techniques is provided in industry with no appraisal of its effectiveness is quite remarkable in an age where most training has to be justified on cost.

To make matters worse, it is quite obvious that in many

professions (e.g. in nursing) training in manual handling techniques is merely a cosmetic exercise which tries to cover-up the deficiencies of a system which is *intrinsically* unsafe. It is recognized that nurses are being asked every day to move weights which far exceed any known modern criteria for acceptable lifting loads and that they are also asked to carry out their tasks at a variety of workstations which, by their design, induce prolonged static loading of the musculature of the spine. The short-term effects of the latter are also well known to those who have changed the points on their motor vehicles! The present concept of training in such situations is therefore both unsound and futile because it does not address the real source of the problem.

Training should be seen as providing an additional measure of safety to a system which has already been adjudged safe. It should not be considered as an alternative to providing that safe system of work.

9.4.2 Selection and screening

It is tempting to assume that strong individuals will be at less risk of developing disorders of the musculo-skeletal system than will weak ones. The argument in favour of this assumption is put into effect – in the form of preselection screening – by some practitioners. There is now good evidence (Troup, *et al.*, 1987) that strength tests and assessments of handling skills do not prove to be good predictors of who will subsequently develop back disorders. The implication of this is obvious and should destroy some of the current myths. That said, however, it may make good sense to identify individuals with prior histories of severe back disorders, with chronic muscle wasting or clinical abnormalities of their musculo-skeletal systems.

For other disorders, such as tenosynovitis, screening tests (e.g. the Hettinger test) and selection based on muscle strengths, have failed to be accepted as either valid or reliable as tools for preselection.

9.4.3 Ergonomics

It is clear that both the previous strategies are flawed by the lack of attention paid to the actual tasks imposed on individuals at work. Work should be designed such that all reasonably fit and healthy individuals can undertake it without risk of ill health. That, you might indeed point out, is all very well as a concept, but it is impossible to achieve in practice – an argument usually advanced by those who wish to justify taking no action that might help achieve this goal. I would be the first to acknowledge that we may never see the end to our endeavours but this should not deter us from trying to achieve a work environment which strives to accommodate as many individuals as possible within a healthy, comfortable, efficient and safe work system. I would propose that, in order to achieve their aim, an ergonomic approach should head the list of preventive strategies.

Statistics from Sweden provide us with a good idea of the percentage of occupational diseases attributable to unusually strenuous motions or postures which are considered to arise from poor ergonomic design. Figure 9.4 illustrates the relative contribution of each of a number of suspected causes to the overall problem. It is perhaps surprising to note just how much greater is the proportion related to poor ergonomic factors over the more traditionally recognized causes. It is also the latter which usually receive more attention in occupational health research.

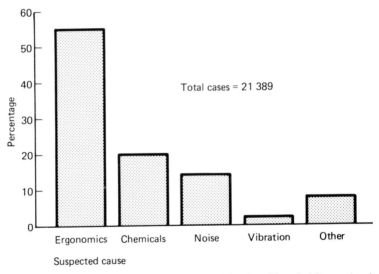

Figure 9.4 Occupational diseases in Sweden, 1983 (National Board of Occupational Safety and Health, Sweden)

The quality of the Swedish data is such that the proportion of the diseases attributed to poor ergonomic factors can be subdivided to indicate parts of the body affected and the relative distribution between the sexes. It is not clear why such strong sex differences exist with respect to the part of the body affected

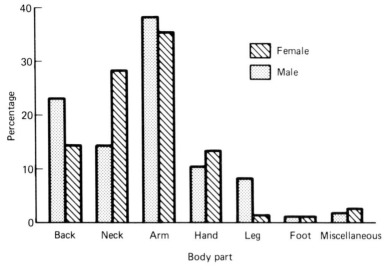

Figure 9.5 Ergonomic diseases by body part and sex (National Board of Occupational Safety and Health, Sweden)

(Figure 9.5). It seèms likely, however, that, rather than being an innate predisposition according to sex, it is more a reflection of the types of work opportunities existing for females.

In order to judge the effectiveness of an ergonomics approach, we must turn to studies where ergonomic intervention has been applied at the workplace.

9.5 Intervention studies

Intervention studies generally take one of two forms: short-term and longer-term.

9.5.1 Short-term studies

The short-term studies are those in which the dependent variables are not necessarily specific disorders but rather manifestations of, for example, chronic fatigue, discomfort or excessive biomechanical load. Some of the many examples of successful intervention are detailed below. All are well documented in the scientific literature.

A recent study of visual display unit (VDU) terminal users by Shute and Starr (1984) considered the beneficial effects on health brought about by the use of adjustable chairs and tables. Subjects reported less frequent neck discomfort, fewer incidents of sore eyes, less upper back pain and less ankle/foot discomfort with the adjustable chairs. The adjustable chairs also helped to reduce shoulder and upper back discomfort. When the tables and chairs were used in combination there was a more dramatic reduction in frequency and degree of discomfort. The operators were observed to use the adjustments frequently.

A study of the shape of handles on cutting knives used in a turkey thigh-boning task (Armstrong, *et al.*, 1982) showed that with a simple redesign – from straight to 'pistol' grip – the biomechanical strain on the wrist could be reduced substantially.

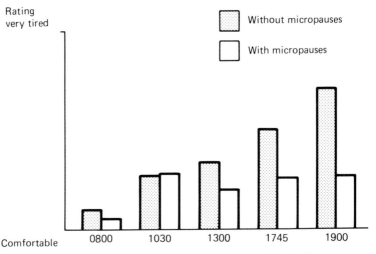

Figure 9.6 Perceived fatigue in the neck during a workday with and without micropauses among data-entry operators. (From: Ehnström, 1981)

Ehnström (1981) showed the effectiveness of taking micropauses of 10 to 15 s every 10 min amongst a group of data-entry operators. The results (Figure 9.6) show how, through this simple expedient, ratings of fatigue were kept at a very low level compared with those recorded on a typical workday.

9.5.2 Longer-term studies

While short-term studies are important for some aspects of health, comfort and efficiency, there are others which have been carried out over longer time-periods and have considered in greater depth the changes in sick-leave attributable to musculo-skeletal disorders, and the possible cost savings which might be made. One such study was carried out in Norway at the STK, Kongsvinger plant. It is described further in Chapter 12 of this book and in detail elsewhere (Westgaard and Aaras, 1984; 1985; Spilling, Eitrheim and Aaras, 1986). A highly significant reduction in sick-leave associated with musculo-skeletal disorders was achieved by following simple ergonomic improvements at the workplace, e.g. the provision of adjustable work surfaces and improved lighting. It was also discovered that job turnover had been reduced and that the cost savings represented a ratio of investment:savings of approximately 1:10. It would seem that the ergonomic improvements had important beneficial effects.

9.6 The problem of 'the case'

Only 1 in every 2500 episodes of back pain in the general population (i.e. 0.04%) ever reaches the stage at which surgical intervention is required. Similarly, hospital referrals probably only represent some 1.5%. The result of this 'tip-of-the-iceberg' phenomenon is that a very different perception of these problems exists in the medical world as compared to that on the shopfloor or by the general public. This bias is seen most clearly in the medical research literature where case studies are frequently reported which do little to shed light on the bulk of the problem. Such reports may be in danger of distorting the general perception of the problem to the detriment of the majority of sufferers and the real needs of the general population.

In a similar context, the recent increase in compensation claims, both in the UK and abroad, has led to a focus on individual cases where, for example, gross mechanical trauma is an obvious contributory factor. This has led too often to the sufferers from back injuries experiencing onset of trauma and with, perhaps, worse prognoses being ignored.

9.6 Conclusions

I have tried to provide a fleeting view of some of the key problems which continue to exist with respect to the quantification and prevention of work-related musculo-skeletal disorders in our society. I have attempted, in so doing, to emphasize the size and cost of the problem and to suggest possible approaches for their prevention.

Ergonomic intervention at the workplace and the concept of designing for the user are central to any such programme. If these are given the support they now warrant, both by implementation of ergonomic principles and through the generation of improved data sources to aid job designers, then we can look forward to the development of workplaces which start to fulfil four essential criteria, namely, that they are healthy, safe, comfortable and efficient.

9.7 Glossary

'Beat' conditions:	chronic, often severe inflammation from continuous pressure on a small area of skin.
Cross-sectional study:	a study of diseases and associated factors as they exist in a defined population at one particular time.
Dynamic muscular work:	work involving changes in muscle length during performance of a task.
Epidemiology:	the study of the distribution, determinants and deterrents of states of health in a population.
Hettinger test:	a test of hand skin temperature before and after exposure to vibration.
Psychomotor:	related or referring to the motor effects of mental processes.
Retrospective study:	a study which tests hypotheses about the disorders by collecting data about the subjects based on experiences and events in their past. These are compared with the same data collected from people without the disorder.
Static muscular work:	work undertaken without a change in muscle length.
Tenosynovitis:	inflammation of the sheath surrounding the tendon.

References

ARMSTRONG, T. J., FOULKE, J. A., JOSEPH, B. and GOLDSTEIN, S. (1982) 'Investigation of cumulative trauma disorders in a poultry processing plant'. *Am. Ind. Hyg. Assoc. J.*, **43**, 103–116

EHNSTRÖM, G. (1981) *Ett forsok med mikropauser bland anstallda vid postgirots bokforing – savdelningar.* Examensarbete, National Board of Occupational Safety and Health, Umea, Sweden

HADLER, N. (1985) 'Illness in the workplace: the challenge of musculo-skeletal symptoms'. *J. Hand Surg.*, **10A**, 451–456

JAYSON, M. (1987) (ed.) *The lumbar spine and back pain.* 3rd edn. Churchill Livingstone, London

KELSEY, J. L. (1982) *Epidemiology of musculo-skeletal disorders,* vol. 3. Oxford University Press, New York

NATIONAL BOARD OF OCCUPATIONAL SAFETY AND HEALTH (1987) *Occupational injuries in Sweden,* 1983. Swedish Work Environment Fund, Stockholm

OFFICE OF HEALTH ECONOMICS (1985) *Back pain.* Office of Health Economics, HMSO, London

PUNNETT, L., FINE, L. and KEYSERLING, W. M. (1987) 'An epidemiologic study of postural risk factors for back disorders in industry'. In: P. Buckle, (ed.) *Musculoskeletal disorders at work.* Taylor and Francis, London

SHUTE, S. J. and STARR, S. S. (1984) 'Effects of adjustable furniture on VDU users'. *Human Factors,* **26**, 157–170

SNOOK, S. H., CAMPANELLI, R. A. and HART, J. W. (1978) 'A study of three preventive approaches to low back injury'. *J. Occupat. Med.,* **20**, 478–481

SPILLING, S., EITRHEIM, J. and AARAS, A. (1986) 'Cost benefit analysis of work environment investment at STK's Telephone Plant at Kongsvinger'. In: E. Corlett, J. Wilson and I. Manenica (eds.) *The ergonomics of working postures.* Taylor and Francis, London, pp. 380–397

TROUP, J. D. G. and EDWARDS, F. C. (1985) *Manual handling – a review paper.* Health and Safety Executive. HMSO, London

TROUP, J. D. G., FOREMAN, T. K., BAXTER, C. E. and BROWN, D. (1987) 'Tests of manual working capacity and the prediction of low back pain'. In: P. Buckle (ed.) *Musculo-skeletal disorders at work.* Taylor and Francis, London

WALLACE, M. and BUCKLE, P. (1987) 'Ergonomic aspects of neck and upper limb disorders'. *Int. Rev. Ergonomics,* **1**, 173–200

WESTGAARD, R. H. and AARAS, A. (1984) 'Postural muscle strain as a causal factor in the development of musculo-skeletal illnesses. *Appl. Ergonomics,* **15**, 162–174

WESTGAARD, R. H. and AARAS, A. (1985) 'The effect of improved workplace design on the development of work-related musculo-skeletal illnesses'. *Appl. Ergonomics,* **16**, 91–97

Chapter 10

The modern office

J. E. Ridd and A. S. Nicholson

The efficiency and productivity of any office is dependent on many different factors, e.g. the expertise and morale of the staff, the quality of management, of data and of the equipment available. Efficiency and productivity are also more generally reliant on the resources of the parent company. However, important though these factors are, an approach that pays attention only to these points is inadequate. We will attempt to demonstrate that the additional consideration of ergonomic principles, as applied to the design of the office system as a whole, can both improve work efficiency and safety, and also reduce the occurrence of associated occupational ill health.

The perennial requirement to improve efficiency levels has previously been tackled largely by implementing the results of studies such as 'operations and methods' and 'work study'. While this approach has been able to achieve some measure of success in the areas of work rate and work output, it may be responsible to some degree for increases in reported dissatisfaction, discomfort and physical disorders amongst office personnel, e.g. as a consequence of adopting awkward postures for prolonged periods. It is clear, therefore, that the system as a whole must be examined in order to increase work efficiency, and not just the competency of staff in their utilization of time and equipment.

It is recognized, of course, that the introduction of electronic office systems in recent years has created major changes in the execution of normal office work. This has also resulted in the appearance of many problems, some foreseen others not, despite the numerous benefits, and it is these difficulties that concern us in this chapter.

The introduction of the new technology, originally viewed as a means of releasing staff from the more tedious office tasks, has not resulted in the much-heralded creation of 'spare time'. The enormous capacity and flexibility now available from even 'stand-alone' systems have opened up new avenues of work that rapidly consume any liberated time and most importantly, in the context of the modern office, also require staff to interact more closely, and for longer periods, with their office systems than ever before. This has both created new problems and contributed to the increases in injuries and unease mentioned in Chapter 9.

The specialized staff skills and the growing complexities of the equipment at their disposal are increasingly important factors in the efficient operation of the modern office; however, there is an equally important and parallel requirement, namely that of the efficient use of such equipment. This subject is rarely given the attention it deserves, and yet the high cost of most electronic systems increases the impact of any mismatch or inefficiency in their operation. While training in the correct use of office equipment is an important element, it cannot readily overcome all of the problems; by addressing more fundamental issues such as the better design of the work, the work tools and of the workspace and the opportunities for operator error, inefficiency or system abuse are reduced considerably. While recognizing that these design considerations can be only complementary to, and not a substitute for, the expertise of staff, it should be realized that if the expertise and time are being utilized, albeit unconsciously, in overcoming system or workplace deficiencies, then these same skills cannot be available for productive work. The other factors are vitally and equally important and must be considered if the health, safety and morale of personnel are to be maintained, as well as the efficiency and productivity of the office.

These 'other factors' may be brought together under the general heading of ergonomic principles. It is the proper application of ergonomics both to the design and operation of the physical layout of the office and to the scheduling of work that offers the greatest potential for the effective improvement in standards of health and safety as well as in productivity.

Other chapters of this book deal with the consequences of ignoring ergonomic design, e.g. Chapter 4 considers the potential for catastrophy from human error in process control, Chapter 9 outlines the risks of personal discomfort and injury from repetitive work tasks and Chapter 12 provides data for lost time attributable to poor ergonomic design. The office environment can provide as many examples of these consequences as any other branch of business or industry.

This chapter will attempt to present the arguments in favour of the application of ergonomic principles to the office environment. It should, however, be noted that recommendations for the design, construction and operation of ideal workstations, and for the correct environmental conditions, are provided in detail elsewhere (e.g. Grandjean, 1987; Pheasant, 1987) and it is not necessary to repeat them here; the intention is more to explain the reasons for the need for such detailed consideration of workplace design and to offer occasional suggestions for improvement. A cautionary note should be expressed at this point: if this chapter is used as a source of reference for the examination of an existing workplace (or indeed, even, if far more detailed texts are used) it is stressed that the management of such an evaluation is most important. Isolated or ad hoc modifications to individual work aspects may solve a problem in the short term but they may also sow the seeds of discontent for the future.

A simple example may help to demonstrate this point: part of a manual handling task was automated to reduce the physiological cost of heavy manual work. The original system required items to be lifted and carried from delivery vehicles to a receiving counter some 20 m distant. A conveyor belt was provided to assist in this process; items still had to be lifted onto the conveyor and taken off at the other end but the effort involved in carrying the load between these points had been eliminated. The workers at either end of the conveyor were unfortunately then able, and therefore usually required, to work at a much faster rate and the frequency of the peak stress loading to their backs was consequently increased, perhaps leading to earlier onset of back pain than might otherwise have occurred. Clearly, by attending to only one aspect of the process the problem was not resolved but merely transferred to another area: '. . . the efficient running of the plant depends on the proper and complete interaction of the various micro-systems that go to make up the industrial process as a whole' (Ridd, 1986). There are numerous examples within the modern office that parallel this manual-handling scenario. A computer can be installed to remove some of the effort and drudgery from a clerical task; however, the potential of the system allows for an increased throughput and the workers at either end become required to work at the pace of the machine, leading to mental and physical fatigue, probably also to lowered morale and certainly to reduced efficiency.

The same corrective philosophy is directly applicable to the office environment where, equally, the effective management of ergonomic intervention requires a systems approach. This does not mean that the short-term needs have to be ignored – since immediate changes may be effected when necessary – but that their influence on other parts of the work process has to be considered and further modifications implemented where required at a later date. However, if the work system has been designed and installed correctly then it should be possible for it to cope immediately with any short-term change without stressing any of its component parts; essentially, the normal working of the system should be well within its maximum capacity.

In order to examine the problems that occur within the modern office let us consider, in a little more detail, four of the 'micro systems' independently: (1) the work environment; (2) the workstation; (3) the equipment; and (4) the user and the task (Figure 10.1). While each of these will be dealt with separately

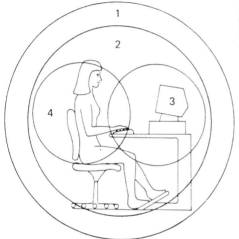

1 Work environment
2 Workstation
3 Equipment
4 User

Figure 10.1 A simple model for the relationship between the components of office work

below, the interactions between the elements must also be considered when evaluating any one of them. The organization of work and the manner in which the user interfaces with the equipment will both influence the task performance, and the design of the workstation itself, and the effect of various environmental factors will also have a bearing on the worker's ability to use that equipment in the manner which had been intended.

10.1 The work environment

The office itself – the architectural features, together with temperature, humidity, lighting, space and air quality – will all have an influence on task performance. Space and layout requirements are changing rapidly with the introduction of more new electronic equipment, and the heat and noise it generates have to be reassessed at regular intervals; lighting requirements may alter with the repositioning of desks and equipment. All these parameters can be measured and the requisite changes made to return the environment to optimum levels (Gill and Ashton, 1982; Ridley, 1983).

One phenomenon to have come to the fore recently but which, however, cannot be quantified easily is the 'sick-building syndrome'. This is characterized by staff reporting a variety of commonplace health conditions such as headaches, sore eyes, sore throats and blocked noses. The problem appears to be confined to modern buildings, or those that exhibit similar characteristics, in particular those that are air-conditioned. It is possible that the heating and ventilation engineers designed a perfectly adequate system, but building and office requirements are never constant, with the numbers of staff and electronic systems operating in any given area changing frequently. The capacity of the system to compensate for these changes may be insufficient, leading either to a build-up of air pollution or fluctuations in the temperature and creation of drafts as the system works to maintain the preset conditions. There is likely to be an increase in discomfort and dissatisfaction in either case leading to reduced concentration and in some cases to lethargy and a general feeling of malaise.

Regular monitoring of the effectiveness of the ventilation system clearly would enable the early identification of potential problems. Appropriate maintenance procedures to reduce recirculation or the introduction of pollutants to the office can pre-empt the onset of the syndrome and, indeed, such attention to staff welfare may contribute to a valuable improvement in worker morale.

Providing a satisfactory work environment, however, may not resolve the problems of all staff, and it is then that we must look to other causes of ill health, e.g. stress-related diseases. The modern office is well able to create stress – from the open-plan style and attendant lack of privacy, to the threat of redundancy – and there are probably as many sources of stress as there are staff employed. A particular concern is the lack of control that an individual has over his or her own task and environment; it is therefore advisable that all aspects of office design should provide flexible systems that enable individuals to create their own 'comfortable' environment, one that essentially ameliorates their particular stress conditions.

Lighting is a common source of complaint in the modern office where it can lead to the adoption of constrained postures causing discomfort (see section 10.2) and to visual fatigue that might manifest itself as itchiness, inflammation or blurred vision, or by referred effects such as headaches. These symptoms are experienced when the visual system is operating to the limit of its capacity. However, simply providing extra general, or even local, lighting may not solve the problem. The work task must be considered as a whole, since some forms of lighting, whilst providing sufficient illumination, may create glare whereby the light source itself or its reflection may cause distraction or irritation. This can usually be overcome by shielding the source, and in many offices this is now accomplished by uplighting.

10.2 The workstation

The workstation can comprise the furniture immediately associated with the work task. For normal office activities such as secretarial/clerical duties and data entry/enquiry, the seated workstation would constitute the chair and desk. The essential design requirements for such furniture might vary according to the manner of occupation of the work station, e.g. (1) is it a single or multi-user position?; (2) is the position to be used for different duties or dedicated to a particular work task?; or (3) if task-related, what special adaptations are required?

The furniture can be set to the requirements of the individual where there is only one user, but where a number of staff are to have access to any one workstation there should be greater flexibility and the capacity for adjustment is essential, otherwise staff may be forced to adopt constrained postures either frequently or for extended periods, and this may lead to discomfort. Evidence from industrial situations would suggest that the cumulative effect of this is an increased likelihood of physical disorder leading to work absence (Fine, Punnet and Keyserling, 1987). This implies that the anthropometrics (body dimensions) of the users should be known in order that furniture with adequate adjustment can be purchased (Figure 10.2). This may appear to be an extra and unnecessary burden, and in some cases, it has to be said, would not solve the problem anyway, since not all manufacturers produce adjustable furniture to the necessary range of adjustment. The purchase of perhaps attractive but inappropriate desks and chairs may well contribute to, rather than ameliorate, the postural problems experienced by the staff. The process of selecting appropriate furniture can be eased by referring to published standards (e.g. Pheasant, 1987) and tables of anthropometric data (e.g. Pheasant, 1986) which provide a basis on which to determine

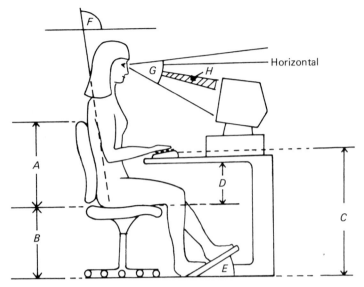

Figure 10.2 Recommended approximate dimensions for office workstations. All dimensions are given to assist the user to adjust the workstation to suit their own preferences (Data from various sources)

A = 50 cm
B = 38–54 cm (adjustable)
C = 70–85 cm (home row on keyboard)
D = $\geqslant$ 17 cm
E = 10–25° (footrest should also be height-adjustable)
F = 90–120° (opinions vary and angle is also task-dependent:
       copy-typing = 90–105°
       interactive VDU terminal work = 100–120°)
G = 35° (cone of easy eye rotation)
H = 5° (normal line of sight)

their acceptability for the working population. The interested reader is also referred to the comprehensive guidance offered by Grandjean, Hunting and Pidermann (1983) and in a publication on ergonomics and computerized offices by Grandjean (1987).

This unfortunately, is not the end of the story; the fact that the required adjustment can be determined and that appropriate furniture can be purchased does not ensure that it will be used correctly – the adjustment mechanisms must be both easy to understand and simple to operate (Figure 10.3). Some adjustment systems have been found to be far too complex for multi-user stations (i.e. different users coming to the workstation cannot easily remember how to alter the settings and therefore don't bother); in addition, the required adjustment forces for some

Figure 10.3 Operating mechanisms for: (a) typist's chair (two controls for all operations); (b) executive's chair (underside of seat). The difference in operational complexity is not a result of the nominal status of these chairs. Complex mechanisms like (b) can be found on typist chairs and equally simple, yet comprehensive, mechanisms as in (a) are employed on some executive chairs (after: Kleberg and Ridd, 1987)

Back rest height adjustment

Adjustment for seat height back rest angle, and angle of seat tilt

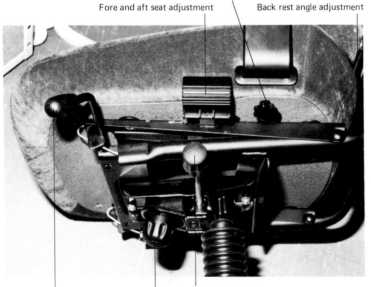

One of four controls for adjustment of arm rests

Fore and aft seat adjustment

Back rest angle adjustment

Seat height adjustment

Hand wheel for tension adjustment of rocking mechanism

Seat tilt adjustment

chairs have been measured to be outside the capacity of some female workers (Kleberg and Ridd, 1987).

It is appreciated that the compilation of a complete list of the anthropometric and physical requirements of the staff is beyond the reasonable capabilities of most organizations, and indeed could not be afforded a high priority; hence, realistically, it is to be hoped that the guidance available can be used as a basis for purchasing decisions and for design. However, one important lesson to be learnt from this is that the brief for the 'purchasing officer' should be far more comprehensive than has been the practice in the past and that, once purchased, it is not sufficient simply for the furniture to be delivered and for the user population to be left to work out how to use it themselves. Even for such apparently mundane equipment as modern desks and chairs, some instruction in the correct manner of adjustment and, indeed, on

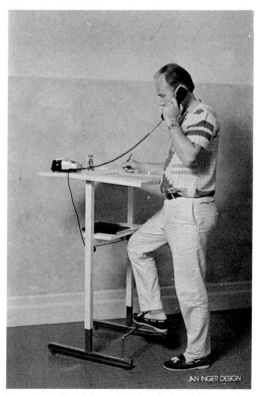

Figure 10.4 Portable workstation designed for use when standing. Note the footrest. (*Courtesy:* Jan Inger Design, Sweden)

the reasons for adjustment must be given if users are to gain the fullest benefit.

The requirement for flexibility is uppermost if a number of tasks are to be carried out at the workstation. The use of fold-away keyboard shelves, or video screens supported on folding, swivel arms may prove beneficial in such situations. But where these are installed, care is needed to ensure that they can be positioned correctly and therefore do not create a situation where their operation requires the adoption of awkward or constrained postures.

Where only one task is normally undertaken, however, a degree of specialization and adaptation should be considered, e.g. 'add-on' shelves to support specialized equipment, inclined desk-tops for reading material, dedicated keyboards, etc.

The use of a footrest should be an available option in any of these situations and the workstation should enable the task to be carried out in either the seated or standing position where appropriate and feasible, thus facilitating postural change; even when a worker is standing, it should be possible to place one foot on a footrest (Figure 10.4).

Where text has to be transcribed (or data entered) from paper to the video display terminal (VDT), some form of copystand should be employed, preferably supporting the paper at eye level. These adaptations and improvements to the workstation also form an introduction to the third area: equipment.

10.3 Equipment

The video screen and the keyboard are the components most routinely considered in this category, but it should not be forgotten that central processing units and disc drives, printers, photocopiers, telex and Fax machines, for example, (not to mention the erstwhile typewriter) present problems both of a similar, and of their own unique nature.

All this equipment produces heat and noise, some items also give off gases and vapours which may contribute adversely to the working environment.

Problems that are component-specific include the following.

1. The video screen should present characters that can be read accurately and without discomfort to the user; there are recommendations and criteria for testing detailed in the (currently) draft British Standard (1987) for VDTs. This,

however, is not solely within the control of the manufacturer as the environment in which the system is installed can affect significantly the clarity of the image.

2. Similarly, criteria for keyboard design are also given in the draft British Standard although this concentrates on the methods of testing. It is suggested here that particular thought be given to the use of detached keyboards with specialized key arrangements for some applications; this should not preclude a consideration of split keyboards where appropriate.

3. The operating instructions for photocopiers (and Fax machines), often represented in abbrieviated or graphic form on illuminated displays, should be clear both visibly and cognitively. (They are manifestly not clear at present. How many attempts are necessary to obtain the correct reduction or enlargement? How many sheets are wasted before both sides of a single sheet are copied correctly?)

4. The paper feed systems for photocopiers, whilst not being perfect, are significantly better than those for most printers and it is to be hoped that designers will soon make use of the current technology to bring these machines up to the high-technology standards of the equipment they support.

In all these examples (none of which are uncommon) the time spent, frustration and annoyance are all taking a toll on the work in hand whether it be in terms of reduced quality or quantity. These problems could be avoided with better design to take account of the user or operator more fully.

We have been concerned so far in this section largely with designing for improved efficiency, but it is now appropriate to consider some of the problems thought to be associated with the use of this equipment. A rather unfortunate and ambiguous name – 'terminal illness' – has been coined to describe a whole gamut of conditions that are thought to be associated with work at VDTs.

Musculo-skeletal complaints, particularly back pain, have been recognized for a long time as being connected with sedentary work, and research has shown that static, seated postures correlate with the onset of low-back pain (Magora, 1972; Adams and Hutton, 1983). The careful design of a work schedule to break-up the time spent in static postures is clearly to be greatly encouraged (see section 10.4).

Another musculo-skeletal condition, commonly called repetitive strain injury (RSI), is thought to be caused by the combination of awkward arm and wrist postures together with, *inter alia,* the speed and frequency of movement that might be associated with the operation of word processors or computer keyboards. Symptoms include pain, fatigue and weakness in the affected limb or area, possibly with local swelling and tenderness (Bammer and Blignault, 1987). As with many of the other alleged health hazards of VDTs discussed below, a cause-and-effect relationship is very difficult to establish; however, recommended treatments for the management of RSI symptoms include rest, change (or rotation) of work tasks, splints, physiotherapy and surgery.

One of the most intense debates that has arisen in connection with the modern office, and in particular in relation to the equipment used, is that of the alleged health hazards of the video monitor itself. Much time and effort has been given in examination and analysis with very little conclusive evidence either to confirm or refute the claims.

It has been claimed that working with a VDT causes health problems ranging from headaches and sore eyes, skin rashes and cataracts, to birth defects and, indeed, specific examples can be quoted of VDT operators suffering from these problems. However, in most cases an equal prevalence of these disorders can be found in populations not working with VDTs. Hence, the cause-and-effect relationship cannot be proven at present.

Studies have considered particularly the radiation emitted from video screens and the US National Academy of Sciences (1983) reported that: '. . . studies have concluded that levels of all types of electromagnetic radiation emitted are below existing occupational and environmental health and safety standard limits of exposure.'

In considering cataracts, Grandjean (1987) reported that all cataractogenic electro-magnetic radiations were lower at VDTs than those derived from normal ambient sources, and that there were no data to support the implied causal association of the video screen.

Tjonn (1984) reported finding cases of reversible facial rashes in a number of VDT operators. He concluded that this was caused by the electrostatic fields created between the screen and the operator in combination with airborne dust particles. However, the cases are few and, so far, only reported in Norway; hence, while the reports should not be ignored it cannot be considered as a serious problem based on the current evidence.

The problem that has created most concern is that of the alleged hazard to pregnant VDT operators. Examples of antenatal problems, spontaneous abortions and birth defects have all been reported in the literature (Berqvist, 1984) but the incidence and demographic variation of these are no different to those found in the population as a whole. The consensus is that there is little or no cause for concern. The potential for harm being caused to the unborn child is said by the Health and Safety Executive (1986) to be unknown, and that: '. . . according to the National Radiological Protection Board emissions from VDTs put neither the operator nor the child at risk, even if the mother is working full time at a VDT.' Nevertheless, any worker who is concerned about this problem should tell her employer since, clearly, any unnecessary anxiety should be avoided during pregnancy.

10.4 The user and the task

The health and safety problems discussed so far in this chapter (with the exception of stress-related ill health) have been created by physical work factors – the air-conditioning, the furniture, and the equipment. Many of the problems experienced by personnel are, however, a result of the manner in which the work is conducted. Worker discomfort, whether it be a headache or a musculo-skeletal problem, is often the result of the required frequency or duration of an activity. Highly repetitive tasks, whether lifting loads or striking keys, can cause wear and tear on the associated musculo-skeletal apparatus. Tasks at the other extreme that require static postures for long periods also increase the liability to injury particularly if any dynamic physical exertion should follow. The users should, therefore, arrange their work to mitigate the effect of such activities, always assuming that those activities cannot be eliminated from the task altogether.

The adoption of a work regime that allows for changes of posture and activity on a regular basis should reduce the risk, or at least delay onset, of discomfort; by reducing the occurrence of such distraction it must be possible to improve concentration and thus raise efficiency.

Some of these postural–activity changes should take the form of rest pauses; their duration need not be long but sufficient to relax the active muscle groups. In many work situations these may take the form of disguised pauses, i.e. where work of a necessary but less taxing nature is carried out as a means of relaxation. Wherever continuous VDT work is undertaken, various recommendations for rest pauses have been made in different countries, but an average break would be 10 min away from the unit for every hour worked (International Labour Organization, 1985). This should

(d)

(c)

(b)

(a)

(e)

Figure 10.5 Examples of exercises designed to relieve muscular tension (after: Kleberg, 1986)

For the neck and shoulders:

(1) Turn your head sideways to your left – and to your right.
(2) Turn your head sideways to your left and put your chin towards your chest twice (a nod) – same to your right.
(3) Lift your left shoulder towards your left ear, and down; the right shoulder towards your right ear, and down; lift both shoulders at the same time towards the ears – rest (shown at (a)).
(4) Lean sideways to your left and to your right with your head and neck relaxed (shown at (b)).

For the back:

(1) Put your hands together behind your neck for support and lean back against the chair backrest (shown in (c)).
(2) Slump forwards relaxed and rise slowly and stretch your arms upwards (shown in (d) and (e)).

For the feet:

(1) Move your feet frequently from toe to heel to increase blood flow in the legs and so prevent swelling of the ankles.

not be viewed as detrimental to work output since, in addition to improving the potential for reduced sickness absence, evidence suggests that the working time lost through rest pauses is compensated by the increased efficiency at other times (Grandjean, 1980). One study proposed the introduction of micropauses for data entry operators (10 to 15 s break every 10 min). The recorded levels for neck fatigue were substantially lower when following this practice than in those work sessions without micropauses (Ehnström, 1981). Kleberg (1986) advocated that some of the 'rest' time for the seated worker should be taken up with light exercises which are designed to prevent, or at least to relieve, muscular tension in the neck, shoulders and back (Figure 10.5).

The manner in which work is scheduled is critical to the health and well-being of the office worker, particularly where the task is monotonous and requires little thought or skill. The introduction of routines that enable task rotation, either between workers or within an individual's own occupation, is to be encouraged. When improvements to the workplace design have been made, according to ergonomic principles, then training in the appropriate exercises for the relief of muscle tension may further ameliorate these problems.

10.5 Conclusion

A satisfactory (ergonomically correct) work environment cannot be achieved simply by installing more lighting, providing a new chair or more sophisticated computers, or by attending, in isolation, to any of the other factors discussed above. These *ALL* have the potential for improving work conditions but – and this is most important – this potential may not be realized unless proper consideration is given to the manner in which each element affects another; indeed, there would be as much scope for creating more problems as for eliminating them unless this condition is honoured. Ergonomic intervention requires a systems approach whereby all these factors are considered, together with their interactions and, where necessary, compromises agreed. The questions raised by the current transformation in office work can be resolved, if tackled properly with far wider benefits than might have been conceived originally.

The arguments in favour of the application of ergonomics within the office environment are strong; the potential benefits are not only represented in terms of reduced ill health and fewer safety problems but also in improved work efficiency and worker morale. It is hoped that the aspects covered in this chapter will help the reader address current conditions with a critical eye, and that the modern office will not, therefore, be an electronic factory fit only for automatons but will evolve into an efficient workplace where staff are not only safe and healthy but are also stimulated by their environment.

References

ADAMS, M. and HUTTON, W. (1983) 'The effect of posture on the fluid content of lumbar intervertebral discs'. *Spine*, **8**, 665–671

BAMMER, G. and BLIGNAULT, I. (1987) 'A review of research on repetitive strain injuries (RSI)'. In: P. W. Buckle, (ed.) *Musculoskeletal disorders at work*. Taylor and Francis, London

BERQVIST, U. O. (1984) 'Video display terminals and health. A technical and medical appraisal of the state of the art'. *Scand. J. Work Environment and Health*, **10**, Supplement 2

BRITISH STANDARDS INSTITUTION (1987) *Draft British Standard recommendations for ergonomics requirements for design and use of visual display terminals (VDTs) in offices*. BSI, Milton Keynes

DAINOFF, M. J. and DAINOFF, M. H. (1987) *A manager's guide to ergonomics in the electronic office*. John Wiley, Chichester

DY, FE JOSEFINA F. (1985) *Visual display units: job content and stress in office work*. International Labour Office, Geneva

EHNSTRÖM, G. (1981) *Ett försök med mikropauser bland anställda vid postgirots bokföring – savdelningar*. Examensarbete, National Board of Occupational Safety and Health, Umea, Sweden

FINE, L., PUNNET, L. and KEYSERLING, W. (1987) 'An epidemiological study of postural risk factors for shoulder disorders in industry'. In: P. W. Buckle (ed.) *Musculoskeletal disorders at work*. Taylor and Francis, London

GILL, F. and ASHTON, I. (1982) *Monitoring the health hazards at work*. Royal Society for the Prevention of Accidents, London

GRANDJEAN, E. (1980) *Fitting the task to the man*. Taylor and Francis, London

GRANDJEAN, E., HUNTING, W. and PIDERMANN, M. (1983) 'VDT workstation design; preferred settings and their effects'. *Human Factors*, **25**, 161–175

GRANDJEAN, E. (1987) *Ergonomics of the computerised office*. Taylor and Francis, London

HEALTH AND SAFETY EXECUTIVE (1986) *Working with VDUs*. HSE Ind. (G) **36(L)**. HMSO, London

INTERNATIONAL LABOUR ORGANIZATION (1985) *Visual display units: job content and stress in office work*. ILO, Geneva

KLEBERG, I. G. (1986) 'Aches and pains in the office'. *Safety Practitioner*, (Aug.), 11–12

KLEBERG, I. G. and RIDD, J. (1987) 'An evaluation of office seating'. In: E. D. Megaw (ed.) *Contemporary ergonomics*. Taylor and Francis, London

MAGORA, A. (1972) 'Investigation of the relationship between low-back pain and occupation. 3. Physical requirements: sitting, standing and weight-lifting'. *Ind. Medicine*, **41**, 5–9

PHEASANT, S. (1986) *Bodyspace, anthropometry, ergonomics and design.* Taylor and Francis, London

PHEASANT, S. (1987) *Ergonomics, standards and guidelines for designers.* British Standards Institution, Milton Keynes.

RIDD, J. E. (1986) 'Ergonomics and manual handling'. *J. Occupat. Health and Safety, Australia and New Zealand,* **2,** 2, 138–142

RIDLEY, J. (1983) *Safety at work.* Butterworths, London

TJONN, H. H. (1984) 'Report of facial rashes among VDU operators in Norway'. In: B. Pearce (ed.) *Health hazards of VDUs.* Wiley, Chichester

US NATIONAL ACADEMY OF SCIENCES (1983) *Video displays, work and vision.* National Academy Press, Washington

Chapter 11

Manufacturing automation

E. N. Corlett

11.1 Introduction

The introduction of the microprocessor into manufacturing has created dramatic changes in the design of equipment and the whole concept of the manufacturing plant. Information technology (IT) has made effective the robot, the flexible and computer-controlled machine and the extensive automation of much of line management. People now talk of the 'automatic factory' as a reality.

It is easy to become immersed in the possibilities of IT and automation, seeing immense opportunities; what can be forgotten is that these systems do not conceive or design themselves. It is true that some of their manufacture is automated, but their installation, maintenance and much of their operation will depend on human decisions and intervention. What is more, their installation and eventual effective use is very much a human problem for, although the man–machine interface (MMI) is more remote from the process than formerly, it is still there. People have to become used to different modes of working, to understand processes and systems in different ways, and are often expected to achieve these changes whilst continuing to produce the company's goods or services.

There are several areas in which it is possible to recognize important effects arising as a result of introducing advanced manufacturing technology (AMT) into a plant. These concern: (1) the consequences of the alterations in the balance of physical and mental activities in work; (2) the direct physiological effects of these changes in activities; (3) the changes in hazards and in levels of risk which can lead to variations in the type and number of accidents; (4) changes in attitudes to work and in work performance as results of changes in the work; and (5) the new jobs which can arise and their long-term effects, of which we have little knowledge. This is not an exclusive list; however, this chapter will deal mainly with these five areas, which are seen as important by many workers dealing with this subject.

11.2 What is advanced manufacturing technology (AMT)?

Before discussing the effects of AMT, and the current and potential contribution of ergonomics, some overview of AMT is required. As mentioned earlier, the microprocessor has been the cause and kernel of advanced manufacturing equipment as well as the advances in manufacturing management. The most widespread evidence of the use of microprocessors is in the use of visual display terminals (VDTs), particularly with the software, which enables them to be used for word processing. At this relatively simple level they are ubiquitous in offices. The display unit and keyboard, or keypad, are widely evident also on the shopfloor. Apart from the free-standing micro-computers or terminals, which are used for data logging and offline programming, computer numerically controlled (CNC) machines as well as robots will now carry their own control computer. For groups of machines, set out as a cell, they will be controlled from a mini computer, or larger machines, via a local area network (LAN) such as manufacturing automation protocol (MAP). This system, developed with major inputs from General Motors, is a network which allows different machines to be controlled by a central system, and to be linked to a computerized draughting system.

The feasibility of such integrated systems has required the development of sophisticated and reliable sensors to identify the positions and conditions of machine parts and components. Furthermore, transport systems for both components and tools have been devised between the machines, and between machines and stores. A component undergoing automated manufacture has to be identified, its orientation discovered and then must be carried to the machine and loaded. It is obviously necessary to check that the previous part is not still in the machine, and that the machine is orientated to receive it. During and after production of the component it has to be checked, sorted and sent to assembly in appropriate quantities. The engineer plans to do this, for all the component parts, so that they arrive at every stage 'just in time' for their ultimate assembly (the 'JIT' control procedure). Assembly is increasingly an automated process, as is subsequent testing, although the progress towards totally automated assembly is still slow and only a relatively few products are built in this way.

Most people think of robots when automation is mentioned, but the major applications for robots are in the handling of parts, e.g. for machine loading and repetitive arc-welding and paint spraying. Specialized robots (e.g. pick-and-place units) are used in particular industries, the aforementioned units being used in the fitting of electronic components to printed circuit boards, for example, or in simple assembly or component transfer tasks.

Computer applications in engineering offices are increasingly common. Many engineering companies now use computer-aided design (CAD). Instead of the drawing-board and square, the draughtsman has a VDT with light-pen and mouse input devices as well as a keyboard. The object is drawn on the screen with these devices, can be rotated, sectioned, stress-analysed and dimensioned. In some cases (and eventually in most cases) the database created during the design of the item can be used to program the machines which will shape it, so the total integration of the manufacturing sequence is possible. This is referred to as CAD/CAM, the last three letters standing for computer-aided manufacture.

The last major step in the automation chain is computer-integrated management (CIM) of which computer-aided production management (CAPM) is a part. The complex timings and sequences to ensure that bought-out parts, and those manufac-

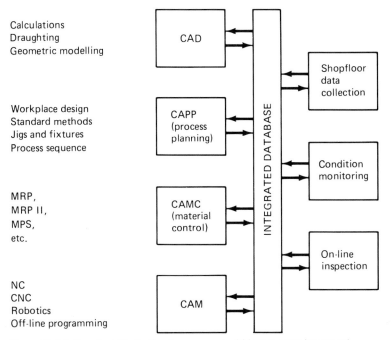

Figure 11.1 A flowchart illustrating the processes within computer-integrated manufacture

tured in-house, are scheduled to arrive at the right place, at the right time and in the right numbers are now commonly done on computers. Much of the data for this comes from the CAD database, but other information comes from sales, purchase departments and those internal operations such as process and capacity planning. Hence, there are important database management needs as well as the planning, simulation and executive software to run the system. Figure 11.1 shows how these various inputs are integrated in a CIM system.

11.3 Health-related factors

Even a superficial observation of many jobs in office or factory 20 or 30 years ago would show that people were active in what they did. Clerical personnel used ledgers and typewriters, etc., visited filing cabinets or filing rooms and moved between offices, frequently carrying things. In factories they lifted material, physically operated machines and controlled the processes directly, e.g. by machining or assembling. There was physical activity and also a direct correspondence between what people did and what the process achieved.

The jobs will be seen to be very different where advanced technologies are used. Keyboards are still in use, and machines still operate, but the users are much less physically active. In the office much can be done just using keyboard and VDT: typing, filing, hard copy or screen-based information transfer to other offices are commonplace. The shopfloor is equally transformed. Machines load themselves, switch themselves on and off, change their own tools and monitor their own performance, where automation is well advanced. Components are assembled into major segments of a product and, in some cases, to the entire product. The operators do not control much of this directly, but by operating the computer which controls the machinery. Furthermore, since direct control is no longer necessary, the operator may supervise several machines or have work covering a number of tasks. Hence, the detailed behaviour of the processes may not be so familiar to these personnel as formerly.

The effects of these changes are most readily seen in offices. Increased use of the VDT leads to long periods of sitting in a limited range of postures, where eyes cover a restricted range of distances and hands and arms are constrained more and more to the keyboard. Reaching for paper and rolling it into the machine, for example, are no longer necessary. All these activities are done while sitting.

The results of these changes can be recognized from the widespread concern with the changing pattern of reported sicknesses. Three of these which caused early concern are the effects of radiation from VDT screens, the occurrence of repetitive strain injury (known as RSI but also reported under different, and in some cases more accurate names, e.g. upper limb strain injury (ULSI)), and anxiety over 'visual fatigue'.

Whatever the faults of earlier VDT screens, it is probably safe to say that no modern apparatus gives off any harmful radiation. Investigations still, nevertheless, continue to assess whether any evidence for long-term effects can be identified. The subject can raise strong arguments at scientific meetings but the balance of evidence, based on work by scientists independent of any involvement with manufacturers, does suggest that this equipment is safe.

Of more concern, because their effects are more evident, are the visual and musculo-skeletal effects. A screen does not as yet have the same contrast or clarity as the printed page. The extended periods of use (e.g. while transferring data from papers in a copyholder to the computer – both viewing points at a similar range, combined with the differences arising from the screen-based image) lead to serious levels of complaint about visual fatigue. An additional problem for some people comes from the static charge present on some screens, which has been linked to skin rashes and eye discomfort.

Although technological developments will reduce or remove some of these causes, the prevention of eye discomfort must include changes in focal point, the opportunities to look at objects at a range of distances and under varying levels of illumination. As with other bodily functions, the eyes benefit from variations in activity, and to constrain their use to long periods with limited focus will lead to long-term damage. This fact is well-known: draughtsmen and clerks, lacemakers as well as miners, have in the past suffered deterioration in visual function as a result of continuous close work or inadequate illumination.

The effects of postural constraint, combined with long periods of use, give rise to the RSI problems. To operate a keyboard requires that the hands, arms, shoulders and neck, as well as the trunk, maintain a close relationship with each other, and with the equipment. Hence, the same muscle groups remain active and the same joints are under load over long periods. However, joints and muscles are designed to be active, and then relaxed, and to experience varying loads from various directions. Treated in the ways which long activities at a terminal require, tendon sheaths suffer inflammation, as do joint capsules. The symptoms are pain and loss of use whilst the cure is generally physiotherapy and rest over a long period.

Other complaints arising from the extended periods of sitting are back troubles and swelling of the legs. The former occurs due to the constraints on trunk position and the often inadequate seating provided. An office seat is a tool for doing a job and cannot be selected independently of the tasks. Although ergonomic workplaces exist, ergonomic seats do not exist independent of the job. Seats should allow as large a range of postures as possible without hindering the work, and allow frequent changes to alter the distribution of loads between muscles and joints. The swelling of the lower legs and feet is a direct consequence of lack of activity and can only be reduced by more exercise.

As with the visual problems, the improvements lie in changes in work tasks. Convenient though it may be for office managers to employ keyboard or data entry operators, their jobs are in direct opposition to what is necessary for health. It has been recommended in some countries, that no more than 2 h should be spent in keyboard work in any half-day period; although other strategies have been recommended, such proposals encourage managers to look at the totality of office tasks so that better work patterns can be generated. The serious incidence of the problems described can thus be reduced while other benefits can be gained in the realms of performance and motivation. These latter will be discussed later in sections 11.4 and 11.6 respectively.

11.4 Adaptation and performance

Let us now turn from the evident health-related changes to some which are more subtle. Computer systems, whether for office or factory, are designed by computer scientists. Usually it is teamwork, with the team carrying office or factory specialists as well as computer specialists. There are several published examples of companies using their own personnel as part of the teams directly involved with the processes being computerized, but it is not usual practice.

One consequence of company staff standing aloof is that it makes it difficult for a user to gain a conception ('mental model') of how the system works in a situation in which the increased distance from the 'sharp end' of the operation and a computer system whose workings are combined and invisible. In order to use anything we must have understanding. Our use of things is inefficient, damaging or dangerous without it. We will build these models as a result of our own practical experience if we have no other sources, i.e. we will learn from experience. Experience may be the best teacher, according to the proverb, but experience doesn't carry the responsibility or pay the bills for the damage. Furthermore, if we all learned only by experience we would all have to invent the wheel afresh; progress would be limited.

The manufacturers of computer systems are aware of these problems and much work goes on to overcome them. The MMI is a major area of design research, to improve the information transfer, whilst the need for 'transparency' in computer programs (i.e. that one should be able to see easily what they do) is also an important study area.

Some of these problems can be reduced by the involvement of those now doing the work in the design of the new system. This involves investment in people by the company which is unpopular with finance directors, who view people as a cost, not a resource. Nevertheless, this investment would maintain the current momentum of a company's business whilst providing time for those running the business to concentrate on the new system. There is then some chance that this new system will link with a company's own system in ways which will help it to be brought into use more easily. As a result, people will have a better understanding of what it can do.

A study of the introduction of computers into banking illuminates this point. The system was based on banking practices. As a result, older bank clerks found little difficulty in adapting. They had gone through such stages as comptometers and calculating machines, and the introduction of computers was just another aid to banking which they understood. Younger staff, however, had greater difficulties. They saw the new system as a computer system, knew less about the banking system and, hence, did not link the two so readily. They found themselves less able to exploit it quickly because their mental model was inappropriate and incomplete.

Much information about computer systems is documented in printed manuals and it is common to find these present great difficulties to users. They are compiled by the computer specialists, whilst the system itself is being programmed, and tend to fulfil the specialist's, rather than the naive user's, needs. User-friendliness is not guaranteed by informal greetings on the screen, nor by cartoon presentations. It is estimated by some researchers that it can be as long as a year before a CAD system is fully utilized and users feel fluent in its use. Such periods can be shortened if efficient training is given, but a 5-day training course is scarcely sufficient. As with the office system, the new user is developing a new mental balance, or set, by taking aboard a new system and rejigging the old to develop what should, ultimately, be a system capable of superior performance to that which existed in the past. It is naive to assume that performance will not suffer while such a complex reorientation is in process.

11.5 Accidents and failures

It is tempting to think that, as automation increases, the number of accidents and injuries will fall. In terms of human injury, this may eventually be so, brought about by a combination of fewer people in hazardous areas and less-close involvement with the process. It must not be forgotten, however, that several people have been killed and many injured by robots already. The largest number of these arose during robot programming, where the worker uses a control box to run the machine through its paces. An error by a conventional machine during such an activity, whilst it may damage the machine, is less likely to injure its operator because the range of machine movements is constrained by its structure. For a robot this is not so; what is more, the structure of the machine gives no guidance to its direction of movement in the way that slideways show the movement of a lathe saddle; so the robot is a more hazardous machine in this respect than a lathe. Hence, it is very important that the control box is designed to present unambiguous controls and information displays to minimize the possibility for error in its use.

As robot numbers increase so, too, could accidents. Work on off-line programming is well advanced and commercial systems exist, so this may counter such an increase. However, in maintenance and in the setting of much automated equipment, awkward work situations, such as too-small work spaces and the need for heavy lifting whilst in awkward postures, still exist. Consideration of the problems of maintenance and setting must have a higher priority in design decisions if these hazards are to be reduced.

It is not only human injury which is important in automated plants. The catastrophic failures of complex systems such as at Chernobyl and Three Mile Island remind us that human error can still lead to disaster. So, indeed, can computer system failure, but this point we shall put to one side.

For a complex system to perform well requires particularly that the operator shall have a correct operational model of the system, a true estimate of the hazards of the system and the continued, unflagging motivation to operate it correctly. The importance of the mental operational model has been raised earlier; here, we have another important reason why such a model must be internalized and readily available to deal with the unusual event. This is not easy if the system is reliable; the expectation of failure is low and the skills to deal with it can be rusty. In modern aviation, where aircraft are flown for most of the time by computers, this problem is handled by requiring all pilots to spend regular periods in a simulator, where malfunctions can be introduced to give pilots practice in dealing with them. Such continuing training may become more necessary. It is not a replacement, however, for good ergonomic design. An assessment of most accidents of the dramatic intensity of those quoted in earlier chapters shows a train of events, differing in their relative contributions in every case, which begins with a design defect, either in the technology or in its ergonomic controllability. When an excessive task demand arises, human error is then likely. If this coincides with other adverse contingencies there can be failures in the system, and some of these will result in a catastrophe. While a 'catastrophe' occurring on an automated production line may not be as drastic in terms of loss of life as in other situations, its economic effects may be equally serious.

11.6 Motivation

The question of motivation has been mentioned on several occasions earlier in this chapter. The effects on people when they are set to work with robots are similar to those caused by the old paced assembly line. The operator works at the robot's pace, all the time the robot works, and of necessity on a repetitive task with limited content. It will be deduced that this work situation is not suitable for people at all.

There has always been the temptation to make the technology do everything it was capable of and then fit people into the spaces. The result has been a lot of dull, repetitive work, totally uninteresting for those engaged in it and, hence, not done very well. It is more than half a century since research demonstrated that better performance (i.e. increased output of higher quality) resulted if people had jobs which they could become involved with. When robots were introduced in the Audi plant in Germany some 4 or 5 years ago, the company devised a policy to create jobs that were not monotonous, or which consisted of the bits left over from automation. They planned to raise the level of qualification of the workforce to increase its flexibility and thus increase plant utilization by reducing downtime. Fault identification and rectification were kept close to the process, and groups of operators, etc. Working in a team shared the qualified and the dull, residual work amongst themselves. The results of the policy enabled the company to increase production while reducing costs without shedding staff; indeed, 2000 more people were employed eventually at the plant.

To be responsible for an expensive machine or complex system can be a stress in itself, so that the design must be optimum in ergonomic terms if good performance is to be achieved. Teamwork can also help to reduce stress by sharing loads and providing assistance when workloads peak, as they inevitably do from time to time. The whole area of work design, the creation of jobs which are suitable for, and acceptable to, people in technical and human terms, is of major concern in advanced industries, and will become more so as manufacturing systems become still more advanced. To operate them will require motivation, dedication and skill at all levels. People will have to see such jobs as attractive when compared with other available careers. To trust the operation of such systems to those who cannot get other work is not a likely recipe for success. Hence, the integration of good work design and ergonomic practices with the technological design would seem the most effective way of ensuring an efficient system when it is built. This rather obvious conclusion is not obvious to all, however. While there are some companies who practise such an approach, they are in a small minority in the UK, being in greater evidence in Sweden, Germany and (in their own way) in Japan, with many good examples in the US. (For more details see the further reading list at the end of this chapter.)

11.7 Conclusion

The benefits of AMT are achieved ultimately as the result of properly utilizing the operators, monitors, maintenance people and technical support staff in the system. If we are to gain those benefits, then the requirements of those people must be built into the system. Some reasons for this, as well as some directions to follow, have been discussed here. The knowledge to do this exists, is the will to apply it present?

Further reading

Studies of the introduction of new technology into a range of industries in several EEC countries will be found in reports by the European Foundation for the Improvement of Living and Working Conditions, Loughlinstown House, Shankill, Co. Dublin. In particular there is a consolidated report entitled *The role of the parties concerned in the introduction of new technologies.*

Bengt Knave, P-G. Widebäck (eds) (1987) *Work with display units,* North Holland Publishing Co., New York, the *Proceedings* of a major conference on VDU use and applications held in Stockholm in 1986, and probably the most extensive current review of the applications of computers in offices, etc.

Buckle, P. W. (ed.) (1987) *Musculoskeletal disorders at work.* Taylor and Francis, London, the *Proceedings* of a conference on posture, held at the Robens Institute of the University of Surrey in 1987, which gives a useful survey of postural research and some recent findings.

The literature on new forms of industrial organization is extensive. The second edition of *Managing and developing new forms of work organisation* published in 1981 by the International Labour Organization in Geneva is a useful overview. Reports from the Work Research Unit at the Advisory Conciliation and Arbitration Service, provide case examples in a wide range of industries as well as some reviews of overseas practice. Publications by the Tavistock Institute extending over the last 40 years present some of the early, formative, material in the field. Typical of several early texts from the Tavistock Institute is: Rice, A. K. (1958) *Productivity and social organisation: the Ahmedabad experiment.* Tavistock Publications, London.

Some recent papers on safety in automation will be found in Noro, K. (ed.) (1987) *Occupational health and safety in automation and robotics.* Taylor and Francis, London, the *Proceedings* of a conference on that subject in Japan.

Chapter 12

Is ergonomics cost-effective?

G. C. Simpson

12.1 Introduction

It is extremely easy for those involved in occupational health and safety to be amazed and become frustrated when others fail to appreciate the simple truth that workers should not be harmed or injured. An extreme example of frustration is shown in the opening remark of a text by Cooper and Smith (1985):

> Work is by its very nature, about violence – to the spirit as well as to the body. It is about ulcers as well as accidents, about shouting matches as well as fistfights, about nervous break-downs as well as kicking the dog around. It is above all (or beneath all), about daily humiliations. To survive the day is triumph enough for the walking wounded among the great many of us.

Few of us use such colourful language and most are less overtly political, yet such a sentiment is common among the safety and health professions. It is, however, counter-productive for it ignores three crucially important realities:

1. The promotion of occupational health and safety is, whether we like it or not, an economic issue.
2. In a period of increasing international commercial competition, spare capital is a rare commodity. All investment decisions are therefore competitive and risky. Investment in the promotion of health and safety is part of the same equation as investment in new plant. As one manager in the manufacturing industry put it: 'What is the point in spending vast sums of money to produce the safest factory in the country if the only people working there are liquidators?
3. There is still a large element in UK management who see health and safety as being largely altruistic. Equally, there is still a tendency among trade unions to be more concerned with cash than safety. Neither group would admit to this, nonetheless an

154

examination of what they do, rather than what they say, will often illustrate the point.

This last point all too frequently becomes an apparently insurmountable frustration to health and safety professionals; however, this is the natural consequence of the failure of the health and safety professionals to address the economic realities of points (1) and (2) above. Few of us would contemplate going to a manager with the statement: 'We have a problem with X, it'll only cost you £50 000 to solve it but it'll be really good for your soul.' It sounds foolish, but that is, in effect, the position if health and safety issues are presented in isolation from any other economic argument.

If we wish to win a decision in favour of investment in health and safety, the argument must be developed in a way which shows that the promotion of the physical health and safety of the workforce can contribute directly and positively to the safety and financial health of the organization. It can be done, and the ergonomic approach is often particularly useful in this context, for it is the only one of the health and safety disciplines which includes performance as well as health and safety in its definition.

In order to win this investment in health and safety, it must be possible to prove: (1) that ergonomic limitations within a workplace are detrimental to the day-to-day running cost of the industry; and (2) that ergonomic change improves the day-to-day profitability. It is necessary to prove both; unfortunately, (2) does not necessarily follow directly from (1), especially as ideally (2) should show a net improvement after the investment in obtaining ergonomic advice and the cost of subsequent change has been taken into account.

12.2 The cost of ergonomic limitations

12.2.1 Back pain

Recent publications (e.g. Zuidema, 1983; David, 1983; Bennett and Passmore, 1984) have clearly and consistently shown that back pain is among the most prevalent of health problems resulting in loss of working time. Moreover, it is extremely widespread, e.g. Lloyd, Gauld and Soutar (1986) showed that there is little difference in the prevalence of back pain between miners and office workers (what differences did occur emerged in the + 40 age-groups). It is also now widely accepted that ergonomic problems of workplace design and work organization can often play a significant part in the aetiology of industrial back pain (Troup, Roantree and Archibald, 1970; Ciriello and Snook, 1978; Snook, Camponelli and Hart, 1978; Chaffin and Andersson, 1984). Placing an exact cost figure on such problems is, of course, extremely difficult; however, a number of estimates have been made. For example, David (1983), although making no financial estimate, indicates the size of the problem by suggesting that solely based on DHSS figures for injury benefit claims, the lost time cost to industry in the UK was between 1 and 2 million man-days per annum. Manstead (1984) added the much more graphic statistic that 6 times as many man-days were lost through back pain during 1982 in UK industry than those which were lost as a result of industrial disputes. It is worth noting the significance of this figure, given the importance placed on strikes, etc. by both politicians and economic analysts when considering the performances of national economies. Manstead estimated the total financial burden of back pain on the UK economy as in excess of £1 billion per annum.

An estimate of the cost implications for a single plant is available from within the British Steel Corporation (BSC) as a result of a BSC/European Coal and Steel Community (ECSC) project carried out during the late 1970s and reported by O'Brien, Fairbank and Daniel (1980). This study recorded a total of 4509 man-days lost over a 2-year period within a single plant in BSC. The authors quote that the cost of this loss in social payments alone was almost £68000. They were also able to calculate an allowance for reduced capacity on return to work and recommended than an addition of 23% of the actual days lost would represent a sensible allowance for reduced performance. This translates to an effective loss in the plant of over 5500 man-days in a 2-year period.

Chan *et al.* (1987) show that absence as a result of back pain can play an influential part in planning decisions and that back pain and its associated absence can incur a cost on existing operations. The ergonomists were asked to advise on the work stations for a new engine assembly line which was still on the drawing board. Initial examination suggested four tasks on the line were likely to require improvement and simulations of these tasks were developed. The simulations showed that in three of them, the workplace design was likely to render the target cycle time for the operation unachievable. In addition, it had been proposed that the new line would incorporate a significant reduction in repair loops over similar previous lines; problems and bottlenecks were therefore inevitable and the rate of return on investment in jeopardy. In two of the tasks, the workplace was such as to increase the probability of musculo-skeletal problems, particularly back pain. An assessment of the financial and manning implications of the potential back problem was then carried out, as discussions with the company had shown back pain to be their largest single cause of absenteeism among assembly workers. It was predicted that the absenteeism from back pain on the new line (70 operators) would be of the order of 50 man-weeks per year. This combined with the failure to meet cycle times and the reduced number of repair loops created a very serious problem. Analysis of the ergonomic assessment showed, however, that ergonomic improvements would influence both the cycle time and back pain difficulties. The improvements proposed were studied further during simulation yielding improvements of between 6 and 16% in cycle time and a predicted reduction in absenteeism of approximately 20%. The company considered that these 'savings' made the line viable; however, the cycle time improvements alone would not have been sufficient without the additional cost burden of carrying 'spare manning'.

12.2.2 Workplace and equipment design

The ergonomic literature is full of examples of poor ergonomics in the design of workplaces, tools, machines, etc. However, the cost implications of these limitations are only rarely addressed. Some examples are quoted below of cost estimates for one health issue – repetitive strain injury (RSI) – and one production issue – the design of mining machinery.

Repetitive strain injury is a generic term coined to cover a variety of disorders of the hand, wrist, elbow and shoulder, such as tendinitis, tenosynovitis, carpal tunnel syndrome, epicondylitis, etc. There is a strong tendency for such injuries to be associated with certain occupations which are characterized by rapid repetitive movements of the hand and arm involving extreme joint positions, particularly where pressure or dexterity are also involved. All of these influential factors are amenable to ergonomic change (Buckle and Baty, 1986) in relation to workspace design, tool design and work schedules.

Steemson (1986) states that over 2000 people per year in the UK receive state compensation for such injuries; however the article adds '. . . the suspicion is that those 2000 represent only the tip of the iceberg'. Compensation costs on RSI claims in Australia have reached $A400 million (O'Grady, 1985). At the individual plant level, Hymorich and Lindholm (1966) recorded 2051 days lost over a 7-year period in an electronics assembly plant which represented 1% of the total available time for that job. Welch (1972) notes one factory in his study which had lost 2000 working days in a single year, and Wehrle (1976) in a study of RSI in an upholstery factory recorded $217000 in 'compensation and outside medical costs' over a 5-year period. Wehrle points out that this figure makes no allowance for reduced productivity. Using Wehrle's figures slightly differently, RSI problems meant that the company involved had to produce and sell products providing an additional profit of $43000 per year simply to replace the drain on resource created by the 'compensation and outside medical costs'. More recently, McKenzie, et al. (1985) quote one telecommunications manufacturing facility with over 1000 man-days lost from RSI among a total workforce of 6600 employees.

A recent study on mobile mining equipment by Chan, Pethick and Collier (1985) gave the cost implications arising from ergonomic limitations in machine design by examining a representative range of development machines used in the UK coal-mining industry. (Such machines are used to drive underground roadways with facilities to remove strata, either by cutting with rotating picks or drilling or blasting, to collect the stone debris and load it out to a conveyor system.) Such multifunctional machines can be extremely large; however, by the very nature of mining, they are expected to operate in relatively confined spaces. Detailed ergonomic evaluations identified a number of limitations, the two that were most consistent across the sample being restrictions to the operator's field of view and poor control layout (caused primarily by the design of the machine). One direct result of the poor sightlines was that a man from the team had to act as a 'spotter' during operation. The spotter took up a position which enabled him to see that which was obscured from the driver's view and direct him by using a mixture of hand and caplamp signals.

Detailed studies enabled Chan, Pethick and Collier to estimate a cycle-time penalty of about 5% as a result of poor sightlines and control layout. This, with the knowledge of the cost of drivage operations, together with estimates of the wage costs incurred in the 'unnecessary' function of spotter (multiplied across all similar machines in the industry) revealed a total cost in excess of £18 million per annum.

12.2.3 Human error and downtime

This is an extremely broad area which is receiving increased attention as more emphasis is placed on the human component of system reliability, but few papers consider the cost implications. Some examples of the published literature are quoted below and cover estimates of the cost implications of ergonomic factors on reductions in plant availability and on the ease of maintenance operations which directly influence the period of downtime.

Much of the work on these topics has concentrated on high-tech process industries, such as nuclear energy and petrochemicals, where the potential environmental hazards are of particular concern. The potential 'economic hazard' is, however, also considerable. Williams (1982) has summarized some approaches to costing the implications of human error resulting in plant failure on the basis that sound ergonomic design can halve the human error rate. Collating data from several sources, he concludes that: 'a reasonable point estimate of unavailability caused by human error during electric power generating is about 0.1%. Calculations by Watson (1981) based on the then current UK operating costs have suggested that a justifiable expenditure to achieve a 0.1% improvement in availability would be about £1 million. On the basis that 70% of human error causing unavailability is likely to be induced by design problems, and 50% are recoverable through the use of sound ergonomics in design, Williams concludes that 0.03% of the capital cost of a plant could be a justified investment in ergonomics simply to reduce unavailability. Although this may seem a small percentage, the enormous capital figures involved in new plant now run to billions, and strongly emphasize the economic potential for investment in ergonomic procedures. For example, a representative figure for the Central Electricity Generating Board's capital cost of a nuclear power station is £1.5 billion, giving the investment on ergonomics to reduce downtime as £450 000.

On the question of problems incurred in maintenance, Seminara and Parsons (1982) reported on a wide-ranging study of ergonomic factors during the maintenance of nuclear power plants. It is possible to begin to appreciate the costs involved from only one of the problems they raise, namely that of access. They concluded that between 30 and 80% of maintenance time was devoted to preparation and that a 30% reduction in total time could be achieved if access were 'ideal or unrestricted'. A similar study in the UK mining industry (Ferguson, et al., 1985) identified significant access problems on almost 50% of the maintenance operations studied. Analysis of the annual reports of several major companies shows that it is not untypical for maintenance costs to involve around 20 to 30% of total operating costs, especially in the heavy industries. The total maintenance cost to a company clearly involves much more than those issues (routine checks, topping-up and on-site repair) likely to benefit from improved access. However, even if the access problem is relevant only to 1% of the total maintenance cost, a 30% saving in time could be enormous, e.g. a 30% saving on 1% of the maintenance cost in British Coal (based on 1983/84) would be worth over £1.5 million.

A number of assumptions and approximations have been made in the estimates quoted above. In the calculation of cost implications, however, such assumptions are inevitable. Furthermore, even if the figures derived were halved arbitrarily, they still indicate that poor ergonomics contributes detrimentally to the day-to-day running costs of industry. There is, therefore, a clear potential for economic benefit to arise from the application of ergonomics. But can such a potential be realized in practice?

12.3 Economic return from ergonomic action

The literature which deals with the economics of ergonomic action on health and safety is, unfortunately, not only sparse, but also rather disparate in terms of the topics addressed. The studies quoted therefore have been structured on whether the relationship between the health and safety problem and economic return was direct or indirect. The direct category is obvious. For example, reducing back pain saves X. The indirect link is a little more devious. The examples quoted show how health and safety improvements can be incorporated in change without health and safety entering the economic argument. Following the brief descriptions of the studies showing direct and indirect effects, a Norwegian study by Spilling, Eitrheim and Aarås (1986), probably the most comprehensive study of the economics of ergonomics yet published, is described in detail.

12.3.1 Direct benefit from ergonomics action on health and safety

Teniswood (1982) described an ergonomic evaluation into back pain in an Australian mining company. The initial evaluation precipitated changes in manual handling methods and drivers' workplaces on mobile plant (covering the influence of layout on posture and the effects of vibration) and a new training programme was introduced emphasizing a greater awareness of the actions and activities which create a 'back risk'. At the end of the 2-year period during which the changes were introduced, lost-time back injuries had been halved. Although Teniswood himself does not mention the financial aspect, another paper (Anon, 1983) in referring to this study states that, as a result of the work, the company's operations were more efficient by $A157 000 per annum.

A now classic study by Tichauer (1978) shows clearly how an ergonomic approach to the design of handtools can directly reduce the incidence of RSI. At the workplace studied by Tichauer, there was a high incidence of RSI problems, particularly tenosynovitis, carpal tunnel syndrome and epicondylitis. The study indicated that several aspects of the job, particularly the workplace layout, tools used and the high rate of repetition, were instrumental in the aetiology of the symptoms. The predominant tool (conventional straight-handled, long-nosed pliers) were considered to be critical as they forced the wrist into a bent position close to the very extreme of the voluntary joint movement range. However, it was apparent that by simply bending and slightly lengthening the handles of the pliers, a much more anatomically acceptable wrist position could be achieved (Figure 12.1). The study then compared the incidence of RSI symptoms using the two plier designs. The results are shown in Table 12.1.

The incidence of symptoms with the new pliers was about 50% of that in the group using conventional ones. Tichauer does not mention any financial implications; however an indication can be obtained from figures quoted in another US paper published at approximately the same time. Wehrle (1976) quotes an average

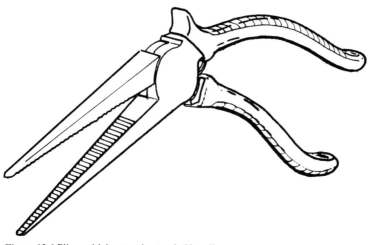

Figure 12.1 Pliers with bent and extended handles

Table 12.1 Incidence of repetitive strain injuries (tenosynovitis, epicondylitis, etc.)
(From study reported by Tichauer (1978) $S = 40$ per group)

Week no.	2	4	6	8	10	12
Straight pliers	7	7	7	8	11	25
Bent pliers	4	4	4	4	4	4

cost of compensation for RSI cases in the US of approximately $3500. Applying this to Tichauer's data (assuming that all cases would have received compensation) suggests a potential saving of over $87 000 arising from a minimal cost change, i.e. the purchase of a new set of pliers for the 40 people involved.

Economic benefit can be shown even when very small numbers of people are involved as work by Drury, *et al.* (1983) indicates. The study concerned five men manually stacking boxes on to a shipping pallet. Ergonomic analysis and redesign reduced the workload to within acceptable physiological limits and achieved a 50% reduction in spinal stress. The financial returns quoted were $374 per man-year through a reduction in lost-time back problems and $3000 to $4000 per man-year in improved efficiency through the workload reduction.

12.3.2 Indirect benefit from ergonomic action

In the first example quoted in this section, the link between health and safety, ergonomics and economics is slightly complex in that, while the ergonomic change produced a significant economic return through a direct effect on production, a potential health problem (extremely adverse thermal conditions) had to be overcome en route. In the other examples quoted, the link is more obvious, albeit indirect.

Feinstein and Crawley (1968) reported a study of the control pulpit of a slab shear operator in the steel industry. The problem was to remove the 'pipe' (which occurred predictably due to the process of cooling) at each end of the slab without removing excess steel. However, the pipe depth was unpredictable and the radiant heat flux from the slab was sufficiently high to require a remote cabin position. The original position was over 10 m away from the shear blades. This distance, together with the angle of view (which meant that in checking the front cut, the slab had to be reversed beyond the cabin position) meant that the operator was forced into one of two equally disadvantageous practices: (1) in an attempt to get an accurate cut, it would be necessary to have several loops to and from the shear blades for both front and rear cuts, thus creating a bottleneck; or (2) to avoid bottlenecks, the operator would deliberately over cut, creating the need to recycle good steel.

The benefits of a cabin position close to the blades were, of course, obvious; however, the problem of heat flux had to be overcome. Following the laboratory studies of the relationship between radiant heat stress and performance by Kettingham (1970), it was shown how the cabin could be sited close to the blades by using gold laminate glass for reducing heat flux to the operator. Additional improvements were also incorporated to improve vision. Data collected by the plant management covering delays due to bottlenecks and the extent of recycling needed both before and after the introduction of the new cabin indicated savings of £120 000 per annum. This related to a capital cost for the new cabin of approximately £10 000; the redesign had a pay-back period of less than 1 month on the capital invested.

A study by Chan, et al. (1985) clearly indicated that a lack of consideration of ergonomic factors in the design of mining machines used to advance underground roadways had both health and safety implications, not only for the drivers, but also for other men in the vicinity. Two problems were paramount: (1) the driver's sight-lines were restricted severely by his position on the machine (an example of the extent of this problem is shown in Figure 12.2); and (2) many of these machines had extremely odd

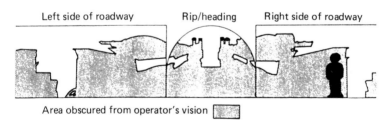

Left side of roadway   Rip/heading   Right side of roadway

Area obscured from operator's vision

Figure 12.2 An extreme example of restricted vision. (Sightlines superimposed on to a heading roadway)

control locations and these, together with the sight-line problem, created a series of very awkward driving positions, as shown in Figure 12.3. Some of the postures, involving twisting, leaning, reaching, etc. are those which are known to increase the risk of chronic musculo-skeletal problems, especially if they are repetitive.

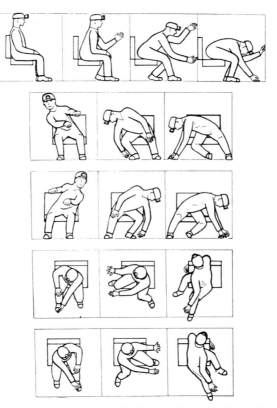

Figure 12.3 Postures adopted during machine operation. (After Chan, *et al.*, 1985)

The implications in terms of health and safety for others working near, or passing, the machine arise primarily from the sight-line problem. First, the inability to see a colleague from the driver's position is seen clearly from Figure 12.2. The second implication is more subtle. In order to overcome the sight-line problem, a man from the development team is deployed during machine operation to act as a 'spotter'. The spotter's function is to take up a position, usually close to the front of the machine, from which he can see clearly the cutting head or drill tips. Examples of some of the spotter positions are shown in Figure 12.4.

Such positions increase the physical danger, for the machines are track-mounted and operate in confined spaces, making them difficult to manoeuvre. The environmental hazards are also increased, for the spotter's position is much closer to the noise and dust source than is that of the driver.

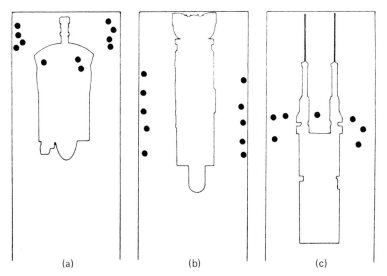

Figure 12.4 Spotters' positions for three machine families: (a) cutter blades; (b) continuous miners; (c) drill loaders

The *potential* health and safety hazards arising from lack of ergonomic thought in the design of such machines were, therefore, obvious. However, fortunately for the men, no evidence could be found to indicate that the potential hazards were reflected in actual health and safety problems. This created a considerable problem in relation to promoting design improvements and an alternative route had to be found. Chan, *et al.* achieved this by examining the operational cost incurred. As a result of this and similar arguments, ergonomic design manuals (produced by Chan, *et al.*) are being distributed to all designers of such equipment. Through the use of these manuals, the health and safety problems will be ameliorated even though they were never used in the argument to promote change.

12.3.3 Action in ergonomics at STK, Kongsvinger, Norway

The study reported by Spilling, Eitrheim and Aarås (1986) covering ergonomic improvements to reduce musculo-skeletal problems in STK's telephone plant at Kongsvinger is probably the most extensive economic assessment of ergonomic design yet published.

The plant is concerned primarily with the assembly and wiring of electro-mechanical components on to telephone switching panels.

The predominantly female workforce, prior to 1974 worked at individual stations with panels at fixed work heights. These workstations '. . . created excessive muscular load due to the need to adopt awkward postures at work' and this was reflected in sickness absence with 5.3% of available work time lost through sick-leave for musculoskeletal conditions.

An extensive redesign of all the workstations was carried out during 1975, using ergonomic principles. A particular emphasis was placed on the need to give each operator greater flexibility allowing, for example, sufficient adjustment to work in both seated and standing positions. Numerous other changes were introduced such as improved seating, tilting worktops (to shorten viewing distance without neck-bending) improved tools, etc. Major changes were also made in lighting and ventilation.

The first level of analysis was simply to check whether the ergonomic change had influenced sickness absence, labour turnover, etc. Figure 12.5 shows the analysis of sickness absence data. In the period 1967–1974 (i.e. before ergonomic intervention)

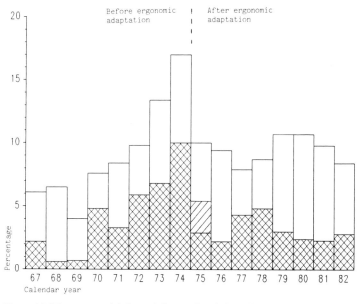

Figure 12.5 Long-term sick-leave (of more than 3 days' duration as a percentage of possible working time each year) at STK, Kongsvinger, from 1967 to 1986. The hatched parts of the columns indicate long-term sick-leave due to musculoskeletal illness. Single hatching in 1975 indicates musculo-skeletal sick-leave beginning in 1974. (From Spilling, Eitrheim and Aarås, 1986)

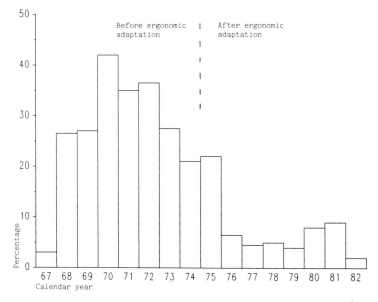

Figure 12.6 Labour turnover (percentage of average numbers of workers each year) at STK, Kongsvinger from 1967 to 1986. (From Spilling, Eitrheim and Aarås, 1986)

musculo-skeletal absence accounted for, on average, 5.3% of the production time available. In contrast, the average level for the period 1975–1982 was 3.1%, a difference which was statistically significant. However, in the period prior to change, the trend was upwards with the figures for 1973 and 1974 being 6.8 and 10% respectively although the average was 5.3%. Between 1979 and 1982, however, the level of absence was almost static at just below 3%. These figures clearly indicate a benefit which has had a lasting effect.

The analysis of labour turnover was even more startling (Figure 12.6). Turnover was running at an average of approximately 30% prior to 1975, whereas in the period 1975 to 1982 it had reduced to slightly over 7.5%. Labour turnover could have been influenced by a whole gamut of factors which had nothing to do with the ergonomic change; however, during interviews with the work-force, improved working conditions were the most frequently reported factor in the reduction of turnover.

Interestingly, not every measure improved. Short-term sickness absence stayed almost static. While these figures show clearly that improvements had occurred which suggest financial benefit in themselves, they do not, of course, prove such benefit.

A long and extremely detailed financial analysis was carried out covering the direct cost of the changes proposed by the ergonomics study (although the costs of the study itself were not incorporated); all were normalized and appropriately amortized (on the basis of a 12-year life). Savings in terms of recruitment, training, instructors' salaries and sickness payments were also calculated using standard accountancy procedures. The savings totalled slightly under NKr 3.25 million on an investment of approximately NKr 340 000. This gave a present value on the investment in excess of NKr 2.8 million.

12.4 Conclusions

The examples quoted show clearly that investment in ergonomic change to promote health and safety can clearly return a financial benefit., sometimes a considerable one. It is also clear that an ergonomics approach allows the promotion of health and safety via an operational cost argument.

The suggestion made at the beginning of this chapter, namely that improving the physical health and safety of the workforce, thereby also improves the financial health and safety of the company, is therefore tenable and should be used more extensively.

Some may still argue that linking health and safety with economics is wrong, believing that the moral argument is sufficient. Such a position may be justifiable. It is, however, also idealistic. In the final analysis, the economic argument is simply a different means to the same end, improving health and safety. If it is, as it appears to be in these ergonomic studies, a powerful ally in the promotion of change, it should surely become the rule rather than, at present, the extremely rare exception.

References

ANON (1983) 'The human factor', *Mining Mag.*, (Dec.), 15

BENNETT, J. D. and PASSMORE, D. L. (1984) 'Days lost from work due to injuries in US underground bituminous coal mines 1975–1982. '*J. Occup. Accidents*, **5**, 265

BUCKLE, P. W. and BATY, D. (1986) 'Ergonomic aspects of tenosynovitis and carpel tunnel syndrome in production-line workers'. In: D. Oborne, (ed.). *Contemporary ergonomics '86.* Taylor and Francis, London

CHAFFIN, D. B. and ANDERSSON, G. (1984) *Occupational biomechanics.* Wiley, New York

CHAN, W. L., PETHICK, A. J., COLLIER, S. G., MASON, S., GRAVELING, R. A., RUSHWORTH, A. M. and SIMPSON, G. C. (1985) *Ergonomic principles in the design of underground development machines.* Institute of Occupational Medicine No. TM/85/11 on CEC Contract 7247/12/007. IOM, Edinburgh

CHAN, W. L., PETHICK, A. J. and GRAVES, R. J. (1987) 'Ergonomic implications in the design of an engine assembly line'. In: E. D. Megaw, (ed.) *Contemporary ergonomics '87.* Taylor and Francis, London

CIRIELLO, V. M. and SNOOK, S. H. (1978) 'The effects of size, distance, height and frequency on manual handling performance'. In: *Proceedings, 22nd annual meeting of the Human Factors Society*

COOPER, C. L. and SMITH, M. J. (eds.) (1985) *Job stress and blue collar work.* Wiley, Chichester

DAVID, G. C. (1983) 'UK national statistics on handling accidents and lumbar injuries at work'. In: *Proceedings, CEC seminar on prevention of low back pain.* Commission of the European Communities, Luxemburg.

DRURY, C. G., ROBERTS, D. P., HANSGEN, R. and BAYMAN, J. R. (1985) 'Evaluation of a palletizing aid'. *Appl. Ergonomics,* **14,** 242

FEINSTEIN, J. and CRAWLEY, J. E. (1968) *The ergonomic design of a slab shear pulpit at Colvillies Limited.* British Steel Corporation/British Iron and Steel Research Association Report OR/HF/8/68. BSC/BISRA, London

FERGUSON, C. A., MASON, S., COLLIER, S. G., GOLDING, D., GRAVELING, R. A., MORRIS, L. A., PETHICK, A. J. and SIMPSON, G. C. (1985) *The ergonomics of the maintenance of mining equipment.* Institute of Occupational Medicine Report No. TM/85/12 on CEC Contract 7247/12/008. IOM. Edinburgh

HYMORICH, L. and LINDHOLM, M. (1966) 'Hand, wrist and forearm injuries'. *J. Occupat. Med.,* **8,** 573

KETTINGHAM, P. J. A. (1970) 'Performance degradation due to radiant heat stress'. In: *Proceedings, symposium on ergonomics and physical environmental factors.* International Labour Organization Occupational Safety and Health, Series 21. ILO, Geneva

LLOYD, M. H., GAULD, S. and SOUTAR, C. A. (1986) 'Epidemiologic study of back pain in miners and office workers. *Spine,* **11,** 136

MANSTEAD, S. K. (1984) 'The work of the Back Pain Association: past, present and future'. In: J. Brothwood, (ed.). *Occupational aspects of back disorders.* Society of Occupational Medicine, British Medical Association, London

McKENZIE, F., STORMENT, J., HOOK, P. VAN and ARMSTRONG, T. J. (1985) 'A program for control of repetitive trauma disorders associated with hand tool operations in a telecommunications manufacturing facility'. *J. Am. Indust. Hygiene Assoc.,* **46,** 674

O'BRIEN, J. P., FAIRBANK, J. C. T. and DANIEL, J. W. (1980) *Low back pain in the steel industry.* European Coal and Steel Community Report on CEC Contract 7245/50/001. ECSC, Luxemburg

O'GRADY, C. (1985) 'Clockwork jobs plague high tech office workers'. *Computing Mag.,* (Sept.), 12

SEMINARA, J. L. and PARSONS, S. O. (1982) 'Nuclear power plant maintainability'. *Appl. Ergonomics,* **13,** 117

SNOOK, S. H., CAMPONELLI, R. A. and HART, J. W. (1978) 'A study of three preventive approaches to low back pain injury'. *J. Occupat. Med.,* **20,** 478

SPILLING, S., EITRHEIM, J. and AARÅS A. (1986) 'Cost-benefit analysis of work environment – investment at STK's telephone plant at Kongsvinger'. In: N. Corlett, S. Wilson and I. Manenica (eds.) *The ergonomics of working postures.* Taylor and Francis, London

STEEMSON, J. (1986) *Up in arms.* Royal Society for the Prevention of Accidents, Birmingham

TENISWOOD, C. (1982) 'Back injury prevention in metaliferous mining operations. In: R. Rawling, (ed.). *Ergonomics and occupational health*. Ergonomics Society of Australia and New Zealand, Melbourne

TICHAUER, E. R. (1978) *The biomechanical basis of ergonomics*. Wiley Interscience, New York

TROUP, J. D. G., ROANTREE, W. B. and ARCHIBALD, R. (1970) 'Industry and low back problems. *New Scientist*, **45,** 65

WATSON, I. A. (1981) 'Cost basis for availability target setting'. *Terotechnica,* **2,** 121

WEHRLE, J. H. (1976) 'Chronic wrist injuries associated with repetitive hand motions in industry'. Ann Arbor Center for Ergonomics, MSc thesis, University of Michigan

WELCH, R. (1972) 'The cause of tenosynovitis in industry'. *Ind. Med.,* **41,** 161

WILLIAMS, J. C. (1982) 'Cost effectiveness of human factors: recommendations in relation to equipment design'. In: *Design '82*. Institute of Chemical Engineering, Publication Series 22; p. 174. ICE, Rugby

ZUIDEMA, M. (1983) 'Back pain – national statistics in The Netherlands'. In: *Proceedings, CEC seminar on prevention of low back pain*. Commission of the European Communities, Luxemburg

Index